GLOBAL EFFECTS OF ENVIRONMENTAL POLLUTION

South and Central America, the Atlantic Ocean, and portions of Africa and North America are seen on this photograph from NASA's Applications Technology Satellite III, taken on November 19, 1967. Satellites have the capability of monitoring many kinds of pollution on a world-wide basis.

GLOBAL EFFECTS OF ENVIRONMENTAL POLLUTION

A SYMPOSIUM ORGANIZED BY THE
AMERICAN ASSOCIATION FOR THE ADVANCEMENT OF SCIENCE
HELD IN DALLAS, TEXAS, DECEMBER 1968

Edited by

S. FRED SINGER

D. REIDEL PUBLISHING COMPANY

DORDRECHT-HOLLAND

Library of Congress Catalog Card Number 78–118129

SBN Number 90 277 / 0151 2

ISBN-13: 978-94-010-3292-6 e-ISBN-13: 978-94-010-3290-2
DOI: 10.1007/978-94-010-3290-2

FOREWORD

The Symposium on the Global Effects of Environmental Pollution has performed an important task; it has helped to determine the world-wide impact of certain types of local pollution and has uncovered certain unsuspected effects that might hold dangerous implications for the future. This Symposium should help to make the world aware of a crisis that is becoming more ominous and that involves the developing as well as the developed countries – the crisis of the human environment. The causes of this crisis are not difficult to discern. There has been an unprecedented increase in the world's population, an ever-increasing rate of urbanization, and in many countries, a continuous process of industrialization.

Essentially, advancing technology has made it possible for a minority of mankind to achieve affluence and holds out hope for improving the well-being of the great majority. But, because it has not been integrated into the natural environment, this very technology – in industry, in agriculture or in transport – is having many undesirable and potentially catastrophic consequences. Our air, our water and our soil are in grave danger. Many species of animal and plant life have become extinct or are facing extinction. The loss to mankind is grave and even the future of life on earth may be in danger. The challenge is to find ways of repairing the harm already done and to prevent further harm.

The conclusion of the Symposium showed that the problem is not insoluble, but that sustained effort, locally, nationally and at the international level can reverse, or at least stop the trend.

The United Nations Conference on the Human Environment, to be held in Sweden in 1972, should help both developed and developing countries to protect and improve the human environment without relaxing their efforts to further their economic and social development.

Man has the ingenuity and is acquiring the knowledge so save his environment. This conviction is strengthened by the conclusions of this Symposium.

PHILIPPE DE SEYNES,

Under-Secretary-General
for Economic and Social Affairs
of the United Nations

EDITOR'S PROLOGUE

An increased scale of human activity has brought with it pollution, defined as "an undesirable change in the physical, chemical, or biological characteristics of our air, land, and water that may or will harmfully affect human life or that of any other desirable species, or industrial processes, living conditions, or cultural assets; or that may or will waste or deteriorate our raw material resources".* Under certain circumstances, the natural processes are unable to keep pace with the increase of pollutants, and then serious problems arise, which are usually on a local scale. On occasion, however, pollution effects may persist long enough so that the atmosphere or the ocean circulation may spread them over the whole earth.

One classic example, well-documented, is the rise in the concentration of atmospheric carbon dioxide produced by the intensive burning of fossil fuels during the past few decades. The buffering action of the ocean has not been able to keep pace with the increased rate of production, and we now find the CO_2 content increased by about 10% and still rising. But while there is little argument concerning the existence of such an increase, there is no agreement as to the consequences of an increase on the radiation balance of the earth and, therefore, on world climate. There are now many other examples of worldwide effects: pesticides in the world's oceans, for example. And there may be pollution effects extant which we have not yet recognized or whose consequences we cannot yet ascertain.

It is important therefore to examine the situation at frequent intervals, to determine whether a pollutant released in our environment could have far-ranging effects in the biosphere, and to probe particularly all interconnections in order to expose any weak link in the ecological chain. We know that global oxygen production depends greatly on photosynthesis of oceanic phytoplankton. As Lloyd Berkner and Lauristan Marshall have pointed out a couple of years ago, if the minute but increasing amounts of pesticides in the ocean could affect the phytoplankton, then the worldwide oxygen production might be decreased. Most worrisome are situations in which a triggering action sets off a feedback mechanism that preexists in nature. Such feedback mechanisms are believed to be responsible for major changes in the climate and for the production of the Ice Ages. It behooves us therefore to examine very carefully, and even conservatively, all pollution effects arising from human activities. It is important also that this examination involve scientists from different specialties but with broad interests. Many disciplines must be represented, including geophysics,

* 'Waste Management and Control', Committee on Pollution, National Academy of Sciences/National Research Council, Washington, D.C., 1966.

geology, biochemistry, biology, medicine, and ecology. The subject matter has obvious interest to the general public, to policymakers in the government, and to all who are concerned about the effects of man's activities on the environment.

Problems of environmental pollution are causing increasing concern to many groups, both within and outside of the United States government. In the Federal Government, the Federal Council of Science and Technology has set up a Committee on Environmental Quality which includes representatives of United States government agencies having programs in environmental quality and pollution control. The President's Scientific Advisory Committee has set up a Panel on the Environment. At the highest level of government, President Nixon has established an Environmental Quality Council, and legislation has been introduced for an Office of Environmental Quality, directly responsible to the President. The National Academy of Sciences has been operating an Environmental Studies Board, under which there are organized a number of panels. Recently, the Congress organized a Colloquium to discuss a National Policy for the Environment. The American Association for the Advancement of Science now has a Committee on Environmental Alteration concerned with environmental quality problems, and there are many private organizations for whom pollution effects are becoming of increasing concern. An outstanding example is the Conservation Foundation.

On an international basis, also, the need has become apparent for being alert to worldwide effects. In September 1968, the United Nations Educational, Scientific and Cultural Organization organized a Conference on Man and the Biosphere which led to important recommendations. The Government of Sweden has called on the U.N. General Assembly to conduct a Conference in 1972 on Problems of the Human Environment, and the need for a global network for monitoring environmental quality parameters has been stressed by the International Council of Scientific Unions.

It is in this context that the AAAS asked me to organize a Symposium on Global Effects of Environmental Pollution for their Annual Meeting in Dallas, in December 1968. Its purposes were to discuss, in three sessions, the worldwide effects which may arise from local pollution and try to uncover, if possible, hitherto unsuspected effects which might have serious consequences. The afternoon session of December 26, 1968, dealt with the balance of oxygen and of carbon dioxide in the earth's atmosphere, and with the problems arising from nitrogen compounds in the soil and water on the earth. The second session, in the morning of December 27, was devoted to possible effects on global climate produced by air pollution, and to the problem of toxic wastes discharged into the oceans. The third session, in the afternoon of December 27, was devoted to a panel discussion at which an assessment was made of the urgency of various pollution problems, including a discussion of the public policy aspects. I have invited additional papers to supplement those presented at the Symposium, and have added introductions to each set of papers.

In setting up their Committee on Environmental Alteration, the AAAS Board of Directors leaned on a report, some of whose paragraphs appear to me to be especially relevant to this Symposium volume:

"Man's relation to the environment is surely one of the most important problems facing society today. Yet these changes are still of limited public concern and have been given insufficient attention, especially by natural and social scientists."

"Improved public understanding is essential, for successful methods of preventing great and perhaps irreversible damage to the environment will often require public financing and public acceptance, and may require changes in law or in social customs or institutions."

S. FRED SINGER

Deputy Assistant Secretary
U.S. Department of the Interior
Washington, D.C., U.S.A.

January 1970

TABLE OF CONTENTS

PART I

CHEMICAL BALANCE OF GASES IN THE EARTH'S ATMOSPHERE

INTRODUCTION

Johnson reviews our current thinking on the origin and evolution of the earth's atmosphere. It is now widely accepted that the primeval atmosphere contained no free oxygen, and that appreciable oxygen concentrations developed only because of the evolution of living organisms which, in turn, had to adapt themselves to the environmental changes caused by the oxygen.

Today, as was first pointed out by L. V. Berkner and L. Marshall, the balance between gain and loss of oxygen is extremely close and very fragile. Ecological effects of pollution could well lead to a decrease of oxygen on a global scale. This problem is discussed critically by Sisler who considers also the fate of the other major gases in the atmosphere.

Carbon dioxide plays an important role: It furnishes the carbon for plant growth, thereby supporting all forms of life. It also exercises a moderating effect on the world's climate by retaining some of the heat radiation which would otherwise escape into space. The concentration of CO_2 has increased suddenly and markedly since the beginning of the industrial revolution – because of the rapid burning of coal, gas, and oil. The concentration will continue to increase, in spite of the introduction of nuclear energy. As a result, atmospheric temperatures may increase all over the world.

The calculations by Manabe predict an increase of 0.8 °C, which could have appreciable environmental consequences and change the world's climate. But CO_2 is not the only factor which affects the long-term state of climate; other types of atmospheric pollution play an important role (see Part III pp. 127–174).

In his final paragraph, Manabe points to the necessity of considering the exchange of CO_2 between atmosphere and ocean. As Johnson has pointed out, the ocean contains 60 times the amount of CO_2 in various dissolved forms and acts as a buffer for changes in atmospheric CO_2 concentration. But, as develops from the report of Berger and Libby, the problem may be quite complex, with the exchange rate possibly controlled by an enzyme. This, of course, raises the possibility again that worldwide marine pollution may control the concentrations of atmospheric CO_2 by affecting the organisms which produce the enzyme.

One of the major pollution inputs to the atmosphere from human activities is in the form of carbon monoxide, almost all of it produced by motor cars. The fact that this toxic substance has now built up to very high global levels must be ascribed to the existence of sinks, probably soil bacteria which metabolize carbon monoxide. The problems of the sources, characteristics, and fate of atmospheric carbon monoxide are described by Jaffe.

In the final paper, Robinson and Robbins, consider the sources and fate of natural and man-made pollutants of sulfur and nitrogen.

THE OXYGEN AND CARBON DIOXIDE BALANCE IN THE EARTH'S ATMOSPHERE*

FRANCIS S. JOHNSON

Southwest Center for Advanced Studies, Dallas, Tex., U.S.A.

Abstract. The scarcity of noble gases on earth, compared to cosmic abundance, constitutes powerful evidence that the earth was formed without an atmosphere and that the atmosphere has evolved by release of gases from the earth's interior. These gases consisted mainly of carbon dioxide and water vapor, and they contained no free oxygen. The water vapor has gone mainly to form the oceans, and the carbon dioxide has gone mainly into carbonate rocks, and a minor consituent, nitrogen, remains the principal atmospheric constituent. Oxygen has been released mainly by photosynthesis, which involves the consumption of carbon dioxide. Most photosynthesis does not make a lasting contribution to atmospheric oxygen, because the products of photosynthesis undergo decay and oxidation, consuming as much oxygen as was produced in their production. A small fraction (about 10^{-4}) of the producıs of photosynthesis escapes decay and provides a lasting contribution to atmospheric oxygen. This natural conversion of carbon dioxide from the earth's interior to oxygen is completely overwhelmed today, by a factor near 10^3, by the burning of fossil fuels, and the carbon dioxide content of the atmosphere is therefore increasing at a relatively rapid rate. Though detailed studies are lacking, the possibility exists that world climate may be affected. The risk of a serious perturbation appears small, but the problem is only poorly understood and the confidence level in such a prediction is low.

1. Introduction

The earth's atmosphere is not a residual of an early atmosphere that was formed along with the earth. Instead, it has largely arisen by continuous, though not necessarily uniform, degassing of the earth's interior throughout geological time (Rubey, 1951). The gases released from the earth's interior contained no free oxygen; its development in the atmosphere is a notable occurrence on earth, and it results from the presence of life. The release of gas from the earth's interior has probably resulted from local heating associated with tectonic activity. It appears most likely that the release commenced about 4.5×10^9 years ago and that it was initiated by the capture of the moon by the earth (Singer, 1968).

The release of gas from the earth's interior has been much more massive than the present atmosphere would suggest. First of all, water should be considered as an essentially atmospheric gas. The total quantity of water released from the earth's interior approximates 3×10^5 g cm^{-2} averaged over the earth's surface, and most of this is now found in the oceans, simply because the earth is cool enough so that it has condensed. The next most important constituent has been carbon dioxide, for which the release has been about 5×10^4 g cm^{-2} (Poldervaart, 1955). However, most of this has been removed from the atmosphere by dissolving in water followed by its precipitation from the ocean in the form of calcium carbonate. The quantity remaining

* This research was supported by the National Aeronautics and Space Administration under grant NGL 44-004-001.

in the atmosphere today is 0.45 g cm^{-2}, while the amount in the ocean amounts to 27 g cm^{-2} averaged over the earth's surface, or 60 times as much. Thus, all but a few thousandths of the carbon dioxide released from the earth's interior has been extracted from the atmosphere-ocean system and locked up in geologic deposits. Nitrogen has been a rather minor constituent among the gases released, but it is chemically inert and, unlike the other gases, it has accumulated in the atmosphere. The total release has been close to 10^3 g cm^{-2}, and about 10% of this has been removed and locked up in geologic deposits (Hutchinson, 1954). A fraction 10^{-8} of the atmospheric nitrogen is fixed each year (Donald, 1960) but most of this is returned to the atmosphere within a few years through decay of plant materials.

2. The Oxygen Balance

There are two significant sources of atmospheric oxygen: the photodissociation of water vapor, followed by the escape to space of the hydrogen that is released; and photosynthesis, followed by accumulation of unoxidized deposits of organic materials, mainly carbon.

In the case of the photodissociation of water vapor as a source of oxygen, the hydrogen that is produced must escape to space in order to leave the oxygen as a persistent constituent of the atmosphere. If the hydrogen did not escape from the upper atmosphere, it would eventually become oxidized, using as much oxygen as was released during its formation. Since hydrogen is a light gas, it can escape from the upper atmosphere by virtue of its thermal energy. However, under present conditions at least, the rate of escape is rather slow, not over 10^8 atoms cm^{-2} sec^{-1} (Donahue, 1966). Over geologic time, approximately 10^{17} sec, this amounts to 10^{25} atoms cm^{-2}, or 17 g cm^{-2}. The amount of water which must be dissociated to produce this amount of hydrogen is 150 g cm^{-2}. Unless conditions at some earlier time were very different from those prevailing now, the rate was probably no larger in the past. The limitations on the rate of escape are provided mainly by the rate at which atomic hydrogen diffuses out of the atmosphere from its region of formation, and the rate at which the atoms enter into chemical combination with other constituents of the atmosphere. To increase the escape rate, the atomic hydrogen concentration in the photodissociation region would have to build up above the present values – something that the chemistry might not permit. The low temperature of the upper atmosphere is a secondary factor; it makes the upper atmosphere very dry and reduces the number of photodissociation events undergone by water molecules, but this is not of much significance since the hydrogen atoms mainly recombine into water anyhow. Another secondary factor limiting hydrogen escape is the protection from photodissociation provided by oxygen molecules, which absorb the dissociating radiation. In the absence or near absence of oxygen, carbon dioxide would assume this role. However, if neither gas were present and the rate of photodissociation of water vapor were to increase, hydrogen recombination would also increase and the escape rate would not necessarily be affected very much unless the recombination processes involving hydrogen were in some way inhibited at the same time.

The second source of oxygen is photosynthesis, which of course was not active at the beginning of geologic time. The first life apparently originated over 3×10^9 years ago. Cloud (1968) associates this earliest life with the banded iron formations, assuming that an oxygen receptor was needed to avoid oxygen poisoning of the organism and that it was provided by ferrous ions. The last of these banded iron formations is about 2×10^9 years old, and this probably indicates the time when oxygen and peroxide mediating enzymes arose and permitted green plant photosynthesizers to release free oxygen. Thereafter there was probably a generally increasing rate of oxygen production, and finally some significant accumulation in the atmosphere. Berkner and Marshall (1964) identify the oxygen concentration at the beginning of the Cambrian era 6×10^8 years ago as 1% of the present atmospheric value. At this level of oxygen concentration, organisms change their mechanism of energy release from fermentation to respiration, which is far more efficient. Further, the oxygen concentration in the atmosphere at this point is adequate to form a low level ozone layer. This ozone layer would not be sufficient to completely screen out biocidal radiation in sunlight ($\lambda \approx 2500$ Å), but when augmented by a thin layer of water, it would attenuate the radiation to the point where life could exist. Consequently, the oceans became capable of supporting life almost everywhere at this point in geologic time.

Berkner and Marshall further identify the beginning of the Silurian era 4.2×10^8 years ago as the point in time when the atmospheric oxygen reached 10% of its present concentration. Ozone then provided protection from lethal radiation even over land areas, and life emerged ashore.

Most of the oxygen that has been introduced into the atmosphere by photosynthesis in excess of decay of organic materials has been consumed by oxidation of surface geologic materials. There are two ways to estimate the total quantity of oxygen involved. One is to examine sedimentary rocks and compare their degree of oxidation to that of the igneous rocks that were weathered to form them. The other is to determine the quantity of unoxidized organic remains; these include coal and oil deposits, of course, plus a much greater amount of highly dispersed organic carbon in sedimentary rocks. Neither method is very exact.

The total volume of sedimentary rocks is estimated to be about 4×10^8 km^3, or a mass of about 10^{24} g, which amounts to 2×10^5 g cm^{-2} averaged over the earth's surface (Wickman, 1954). The igneous rocks that weather to produce sedimentary rocks contain about 3.80% ferrous oxide. The sedimentary rocks contain 0.9% ferrous oxide. To produce the increased oxidation, assuming that all of the missing ferrous oxide has been oxidized to ferric, 30×10^{20} g oxygen is required. In addition to this, some other oxidation has occurred. The volatiles that have escaped from the earth's interior include 22×10^{20} g sulfur, 300×10^{20} g chlorine, and 42×10^{20} g nitrogen (Rubey, 1951). The sulfur was probably emitted as H_2S or SO_2, and it has been oxidized to SO_3, utilizing from 44×10^{20} to 11×10^{20} g oxygen, depending upon whether it was emitted as H_2S or SO_2. The nitrogen was probably released either as N_2 or NH_3; in the latter case 84×10^{20} g oxygen would have been required to free the nitrogen. The chlorine was probably emitted as HCl, which would react with metal oxides or hydroxides without consumption of oxygen. Finally, the 2500×10^{20} g carbon dioxide that has been introduced into the atmosphere may have been emitted as carbon monoxide, in which case the amount of oxygen consumed in oxidizing it would have been 900×10^{20} g. Thus, the oxygen required for oxidation of crustal materials and volatiles emitted from the earth's interior probably lies somewhere near 100×10^{20} g if carbon was released as a dioxide, and near 1000×10^{20} g if it was released as a monoxide.

The coal, oil and other hydrocarbon reserves are products of photosynthesis that have not become oxidized. They are estimated to amount to 3×10^{18} g carbon (National Academy of Sciences, 1962), or 0.6 g cm^{-2} averaged over the earth. In addition, the carbon in living matter and undecayed organic material amounts to about 1.5×10^{18} g (Rubey, 1951). However, these represent only a small fraction of the surviving unoxidized products of photosynthesis. The estimates again are highly uncertain, but it is probable that sedimentary rocks contain an average of about 0.5% organic carbon. This amounts to about 10^3 g carbon cm^{-2} averaged over the earth's surface (Dietrich, 1963), or a total of 50×10^{20} g. This amount of stored carbon implies that 150×10^{20} g oxygen has been made available to the atmosphere. This suggests that the total oxygen production beyond the amount consumed in decay of organic materials has been nearer ten times the amount now in the atmosphere (10×10^{20} g) than a hundred times. It also indicates that the gases released from the earth's interior have consisted predominately of carbon dioxide rather than monoxide; otherwise the oxygen source would have had to have been about a factor of six larger than can be justified in terms of fossil organic carbon. However, such an amount is not outside the range of speculation (Rubey, 1951).

Table I summarizes the amounts of the four principal gases introduced into the atmosphere, and the amounts remaining in the atmosphere today.

TABLE I

Atmospheric constituents

	H_2O	CO_2	N_2	O_2
Total supply	1.5×10^{24} g	2.5×10^{23}	5×10^{21}	1.5×10^{22}
Total now in atmosphere	10^{19}	2.25×10^{18}	4×10^{21}	10^{21}
Total supply per unit area	3×10^5 g cm^{-2}	5×10^4	10^3	3×10^3
Total in atmosphere per unit area	2	0.45	8×10^2	2×10^2

Note: There is sixty times as much CO_2 in the ocean than in the atmosphere but only about 1% as much oxygen and nitrogen in the ocean as in the atmosphere.

It is speculative as to how uniform the rate of oxygen production has been over geologic time. We will assume that it has been uniform over a period of 3×10^9 years, in which case the average rate of release has been 5×10^{12} g oxygen yr^{-1} or 10^{-6} g oxygen cm^{-2} yr^{-1}. It is worth while comparing this with the rate of photosynthesis today, which is estimated to produce 3×10^{16} g C yr^{-1}, about half on land and half in the oceans (Leith, 1963; Ryther, 1963; Strickland, 1965). The associated rate of oxygen release is 8×10^{16} g yr^{-1}, over 4 orders of magnitude above the average net rate of oxygen production over geologic time. This implies that only about one part in 10^4 of the products of photosynthesis escape oxidation. However, it is that small part of photosynthesis that provides the real contribution to atmospheric oxygen, and allows oxygen to persist in the atmosphere even in the face of losses to the earth's crust by oxidation of geologic materials.

3. Carbon Dioxide

The carbon dioxide annually consumed in photosynthesis is 11×10^{16} g. However, the rate of release of carbon dioxide into the atmosphere by oxidation of recently grown organic materials matches this within about one part in 10^4. Therefore, the removal of carbon dioxide each year to deposit new fossil carbon amounts to about 10^{13} g. The burning of fossil fuel now releases about 1.5×10^{16} g carbon dioxide yr^{-1}, three orders of magnitude greater than the rate of return to the fossil reservoir. In fact, it is an appreciable fraction ($\sim\frac{1}{7}$) of the carbon dioxide entering the photosynthesis cycle each year, although this is of little significance owing to the relatively short cycle time in living material. It is more important to compare the carbon dioxide released from fossil fuels with the carbon dioxide consumed to produce new fossil carbon each year; these figures are 1.5×10^{16} and 10^{13}, showing that we are depleting this resource over a thousand times faster than it is being renewed. The total reservoir of fossil carbon in forms suitable for exploitation has been estimated to be near 3×10^{18} g. When this resource has been fully expended, the oxygen consumption associated with this will amount to about 6×10^{18} g, or 1.2 g cm^{-2} averaged over the earth, not enough to seriously deplete the atmospheric oxygen reservoir.

TABLE II

Time constants

	Rate	(Atmospheric content)/rate
CO_2		
Photosynthesis	1.1×10^{17} g yr^{-1}	20 yr
Combustion	1.5×10^{16}	150[a]
Fossilization	10^{13}	2×10^5
O_2		
Photosynthesis with no decay	8×10^{16} g yr^{-1}	10^4 yr
Net organic production	5×10^{12}	2×10^8
Photodissociation and hydrogen escape	2×10^{11}	5×10^9
N_2		
Bacterial action	4×10^{13} g yr^{-1}	10^8 yr

[a] Present observed rate of increase of atmospheric CO_2 is 0.7 ppm yr^{-1}, for which time constant is 450 yr. The difference is explained by a portion of the carbon dioxide becoming dissolved in the ocean.

Table II summarizes the time constants for atmospheric gases for some of the source and loss mechanisms. These figures should be used with caution for carbon dioxide because of exchange that can take place between the atmosphere and the ocean, which contains sixty times more carbon dioxide than does the atmosphere.

There has been much speculation on the effect of the massive release of carbon dioxide that is taking place as a result of burning fossil fuel. There has been a clear trend toward higher carbon dioxide content of the atmosphere, with an increase of about 10% since the beginning of this century; the exact increase is uncertain because of inaccuracies in the early measurements. However, a clear upward trend is discernible also in recent years (Pales and Keeling, 1965). An important aspect of the carbon dioxide increase in the atmosphere from fossil fuel combustion is the degree to which it is shared with the oceans. Early estimates suggested that most of the carbon dioxide released by burning fossil fuels had remained in the atmosphere and that no significant portion of it had been shared with the oceans. However, the data of Pales and Keeling are much more reliable than the early data and they indicate an exchange time from the atmosphere to the ocean of about 5 years. Their indicated rate of increase of atmospheric carbon dioxide from 1958 to 1963 is 0.7 ppm yr^{-1}, or 0.5×10^{16} g yr^{-1}, which is just about $\frac{1}{3}$ the rate of release of carbon dioxide from fossil fuel burning during that period. However, the rate of accumulation of carbon dioxide in the atmosphere can also be influenced by other factors, such as long-term temperature trends for the earth, changes in patterns of cultivation, etc.

4. Global Effects

We shall inquire as to the steps in the oxygen-carbon-dioxide cycle that are sensitive to perturbation or instability. For example, about half the carbon dioxide consumption each year in photosynthesis occurs in the ocean, where phytoplankton are the primary producers. The growing spread of pollution has already shown up indirectly in midocean area. DDT dust has been observed far out at sea (Risebrough *et al.*, 1968). Also, sea birds, the Bermuda petrel, have been found to have substantial concentration of chlorinated hydrocarbons in their eggs (Wurster and Wingate, 1968), and this has apparently deleteriously affected their reproductivity. Owing to their food habits – they are at the peak of the oceanic food chain pyramid – the chlorinated hydrocarbons have apparently been obtained from fish, whose food source in turn goes back ultimately to the phytoplankton, which presumably gathered the insecticide from dust settling into the ocean. Similar evidence comes from Clear Lake, California, where water containing less than 0.02 ppm DDT produced plankton containing 5 ppm, and fish near the top of the food chain containing enough insecticide to kill birds that ate them. The sensitivity of phytoplankton to such materials is probably variable, and one cannot immediately discount the possibility that insecticide and herbicide pollution may have widespread and unforeseen effects that could affect the plankton that converts a large proportion of our carbon dioxide to oxygen.

This problem is not alleviated by the fact that the net oxygen production – photosynthesis exceeding respiration and decay – is very small compared to the total production. A few ocean areas with anoxic bottom conditions and a few marshy areas in which peat is forming are presumably the key areas for maintaining our oxygen replenishment on a long term basis, because they are reservoirs in which unoxidized organic materials are accumulating. These limited areas are at least as susceptible to poisoning as are the open oceans. It is a matter of importance to man's future to recognize and preserve these areas.

Another alarming possibility is that of instability in the relationship between atmospheric carbon-dioxide concentration and average world temperature. An increase in carbon dioxide, say from burning of fossil fuels, may increase the hothouse effect on earth and increase the average temperature of the earth a small amount; calculations indicate a factor-of-2 change in atmospheric carbon dioxide while maintaining constant relative humidity would produce a temperature change of 2.4 K (Manabe and Strickler, 1964). Heating the ocean would tend to drive carbon dioxide out of solution into the atmosphere, further increasing the hothouse effect. Or one might imagine the effect going the other way as a result of increasing the flow of nutrients to the oceans. By stimulating the growth of phytoplankton, the rate of removal of carbon dioxide from the atmosphere might be increased, reducing the carbon dioxide content of the atmosphere and, with this, the hothouse effect. This could lead to world-wide cooling and an increase in the proportion of the carbon dioxide stored in the ocean relative to that in the atmosphere, further enhancing the effect. Conceivably, the ice ages could have resulted from such oscillations. Alternatively, they may have been produced by a variation in the rate of emission of carbon dioxide from the earth's interior.

Because of the importance of these problems to man's future we should be very confident of our full understanding of them. At present, the ideas and even the numbers must be regarded as highly conjectural. Catastrophic problems appear to be in prospect for mankind because of the population explosion and its associated pollution explosion. One of the many possible ways in which pollution can precipitate a catastrophe is by upsetting the oxygen balance in our atmosphere. However, the time constant for this is sufficiently long that a destructive course may well have been followed beyond the point of no return before it is recognized.

References

Berkner, L. V. and Marshall, L. C.: 1964, 'The History of Oxygenic Concentration in the Earth's Atmosphere', Discussions of the Faraday Society, No. 37.

Cloud, P. E.: 1968, 'Atmospheric and Hydrospheric Evolution on the Primitive Earth', *Science* **160**, 729–736.

Dietrich, G.: 1963, *General Oceanography*, Interscience-Wiley, New York.

Donahue, T. M.: 1966, 'The Problem of Atomic Hydrogen', *Ann. Geophys.* **22**, 175–188.

Donald, C. M.: 1960, 'The Impact of Cheap Nitrogen', *J. Aust. Inst. Agric. Sci.* **26**, 319–338.

Hutchinson, G. E.: 1954, 'The Biogeochemistry of the Terrestrial Atmosphere', in *The Earth as a Planet* (ed. by G. P. Kuiper), University of Chicago Press, Chicago, pp. 371–433.

Leith, H.: 1963, 'The Role of Vegetation in the Carbon Dioxide Content of the Atmosphere', *J. Geophys. Res.* **68**, 3887–3898.

Manabe, S. and Strickler, R. F.: 1964, 'Thermal Equilibrium of the Atmosphere with a Convective Adjustment', *J. Atmos. Sci.* **21**, 361–385.
National Academy of Sciences, Energy Resources Report 1000-D, 1962.
Pales, J. C. and Keeling, C. D.: 1965, 'The Concentration of Atmospheric Carbon Dioxide in Hawaii', *J. Geophys. Res.* **70**, 6053–6076.
Poldervaart, A.: 1955, 'Chemistry of the Earth's Crust', Crust of the Earth, GSA Special Paper No. 62, 119–144.
Risebrough, R. W., Huggett, R. J., Griffin, J. J., and Goldberg, E. D.: 1968, 'Pesticides: Trans-atlantic Movements in the Northeast Trades', *Science* **159** (3820), 1233–1235.
Rubey, W. W.: 1951, 'Geologic History of Sea Water', *Bull. Geol. Soc. Am.* **62**, 1111–1148.
Ryther, J. H.: 1963, 'Biological Oceanography, Geographic Variations in Productivity', in *The Sea*, Vol. 2 (ed. by M. N. Hill), Interscience, New York, pp. 347–380.
Singer, S. F.: 1968, 'The Origin of the Moon and Geophysical Consequences', *Geophys. J. Roy. Astron. Soc.* **15**, 205–226.
Strickland, J. D. H.: 1965, 'Production of Organic Matter in the Primary Stages of the Marine Food Chain', in *Chemical Oceanography* (ed. by J. P. Riley and G. Skirrow), Academic Press, pp. 477–712.
Wickman, F. E.: 1954, 'The 'Total' Amount of Sediments and the Composition of the 'Average Igneous Rock'', *Geochim. Cosmochim. Acta* **5**, 97–110.
Wurster, C. F. and Wingate, D. B.: 1968, 'DDT Residues and Declining Reproduction in the Bermuda Petrel', *Science* **159** (3818), 979–981.

For Further Reading

1. On geochemical processes, including quantitative estimates of the quantity of elements in various geological regimes:
   Brian Mason, *Principles of Geochemistry*, 2nd ed., John Wiley and Sons, New York, 1964.
2. On many assorted problems in the area of atmospheric evolution:
   P. J. Brancazio and A. G. W. Cameron (eds.), *The Origin and Evolution of Atmospheres and Oceans*, John Wiley and Sons, New York, 1964.
3. On chemical and life processes in the oceans and particularly the articles by E. Steeman Nielson and J. H. Ryther for world-wide rates of photosynthesis:
   M. N. Hill (ed.), *The Sea*, Vol. II, Interscience, New York, 1963.
4. A review article on evolution of planetary atmospheres:
   Francis S. Johnson, 'Origin of Planetary Atmospheres', *Space Science Reviews* **9** (1969), 303–324.

IMPACT OF LAND AND SEA POLLUTION ON THE CHEMICAL STABILITY OF THE ATMOSPHERE

F. D. SISLER

Department of the Interior, Washington, D.C., U.S.A.

Abstract. Gross pollution produced by modern technology and world population growth may precipitate changes in concentration of the major gases in the atmosphere which are essential for life and human welfare. The hydrosphere, in particular the ocean, is in dynamic equilibrium with the atmosphere and is largely responsible for controlling the chemical composition of the latter. The equilibrium processes are biogeochemical as well as physical in nature. An upset in the ecological balance, as typified by photosynthesis and respiration processes of marine plankton, is considered. The possible consequences to atmospheric quality reflected in change of concentration of oxygen, nitrogen, and carbon dioxide are discussed, but a major disruption on a global scale is not apparent now. However, subtle changes may be taking place which may have far-reaching consequences in coming years.

1. Introduction

The chemical composition of the earth's atmosphere is controlled to a large extent by biogeochemical cycles existing on land and in the seas. It is estimated that well over one-half of the natural constituents of the atmosphere originate from, and are in equilibrium with, the hydrosphere, which includes the oceans, estuaries, inland seas, lakes and rivers. These water bodies are now faced with gross pollution from a vast assortment of chemicals. Since some of these chemicals are toxic and cumulative, and since most persistent pollutants discharged on land or to the atmosphere ultimately accumulate in the hydrosphere, there is a distinct possibility that the ecological balance in this region could be upset with serious consequences to the quality of the atmosphere. The oceans, the major portion of the hydrosphere, now can no longer be considered the ultimate sink for all types of modern pollutants, even though they have served as a most efficient septic tank since Cambrian times.

In recent years, several provocative papers, notably those of Berkner and Marshall [1, 2], Commoner [6, 7] and Cole [5], point to dangerous trends resulting from modern technological practices affecting natural chemical balances of and between land, sea, and air, as well as the climate in general. Effects from burning fossil fuels, widespread use of inorganic fertilizers, depletion of forested areas with reduction of photosynthetic potential on land, release of toxic or harmful chemicals from industrial factories and agricultural practices, and increasing population have been stressed. Planned use of nuclear materials for power and for earth moving projects are also considered. With respect to the atmosphere, the life-supporting gases of oxygen, nitrogen, and carbon dioxide appear to be in danger of change in relative composition.

Oxygen and carbon dioxide are the principal gases involved in photosynthesis [35]. Keeling and associates [22, 23, 24, 25] have made extensive studies of the distribution of carbon dioxide in the atmosphere and surface waters. Seasonal variations in CO_2

often relate to photosynthesis activity in the oceans. Since the carbon dioxide concentration in sea water is strongly influenced by pH, this physico-chemical parameter is important in oxygen equilibrium [8, 11, 20, 40].

Gambell and Fisher [14, 15] report acid rainfall along the United States east coast in the vicinity of Chesapeake Bay as low as pH 4.0. Rainfall samples obtained synoptically at a site on St. Thomas Island, representative of Atlantic coastal conditions, reasonably free from pollution, usually gave pH values in excess of 8.0 and never less than 7.6. Apparently, this acid rain has been falling, at least sporadically, since the middle 1950's.

The Chesapeake Bay Institute of Johns Hopkins has been measuring the vertical and horizontal pH distribution throughout Chesapeake Bay since 1949 [19]. Beginning the summer of 1959, their published reports showed distinctly acid conditions in various sections throughout this large estuary. The reports covered the time period 1949–61 and low pH readings, below pH 7 (and reportedly below pH 4 at one station in the lower Bay in 40 and 60 feet of water), were observed to be widely distributed, both horizontally and vertically during the remaining years.

Gambell and Fisher postulated that the acid rain resulted from the distillation of acid surface waters in the vicinity. Sulfur acids were prominent in the collected samples of rain, but other acids may also have contributed to the low pH readings. That the Chesapeake Bay may have been a source of the acid rain has not been established, but the fact that highly buffered estuarine waters, as compared to fresh water, could become acidic seems quite unusual.

Besides acid, modern rain may contain a variety of chemicals which are potentially harmful to living organisms. The heavy metal lead is a case in point. In the past 30+ years, the exhaustion of lead from automobiles to the atmosphere as burned lead tetraethyl has increased linearly from zero to greater than 1.6×10^{11} grams per year [41]. Most of this ends up in the oceans. Lead is toxic to many organisms including man. It is now well known that living organisms in the seas have an amazing ability to concentrate trace elements and chemicals [17]. The continued pollution of the seas with toxic substances such as lead, mercury, refractile chemical pesticides may well pose a threat to human welfare. Goldberg [17] discusses this point in detail in this volume.

2. Oxygen

The oxygen balance of the atmosphere is a major concern. The importance of ocean photosynthesis to oxygen concentration in the atmosphere seems well established. Berkner and Marshall, who have made extensive studies of the genesis of oxygen in the earth's atmosphere, have considered the potential degradation of this life-supporting gas as a consequence of man-made pollution [1, 2]. They consider the present O_2 equilibrium to be unstable. This assumption is based on a rather involved theoretical model of the genesis of the present atmosphere since pre-Cambrian time of -27×10^8 years. Prior to this period, the authors reconstructed the secondary, abiogenic atmosphere from Urey's basic premise of the absence of a primordial atmosphere. The secondary atmosphere evolved from leaching of reducing gases and water vapor from volcanic processes. The rarified secondary atmosphere (0.001% present atmosphere) permitted the penetration of short wavelength ultraviolet light (0.1–0.2μ) causing photodissociation of water vapor to form H, O_1, O_2, and O_3.

Eventually the concentration of O_2 built up to a point where further dissociation of water was restricted because of the Urey 'shadowing' effect. The overlying O_2 filtered out sufficient short UV such that a self-limiting, steady state process prevented further buildup of oxygen.

With the appearance of photosynthetic life, the oxygen level increased sharply in a relatively short time since photosynthesis uses visible light (0.4–0.8μ) which is not filtered out by increasing oxygen concentration.

Berkner and Marshall reason that the present O_2 equilibrium is unstable because a relatively small drop in production (as might be caused by biocides) will result in a sharp drop in concentration. They assume that the oxygen demand by reduced chemicals and respiration is precisely balanced by present earth's oxygen production. Once the scale is tipped by lowered production, the consequent lowered O_2 concentration will permit more lethal UV to reach the earth's surface, further stifling photosynthesis.

The strength of the oxygen model presented by the authors rests heavily on their evaluation of the limiting effects of ultraviolet radiation on photosynthetic plants. An important question concerns the matter of shielding. Ultraviolet at any wavelength is reflected, scattered or absorbed by thin sections of most substances more dense than gases. Many plants as well as animals are protected by an integument which is impervious to UV. The sunbather is protected from mutagenic UV by a thin coating of oil. Atmospheric dust should screen out much of the UV. Pure water, although more transparent to short UV radiation in a rarified atmosphere, rarely occurs in nature. Most ocean and lake waters contain sufficient impurities in the form of suspended and colloidal materials, such that absorption coefficients based on pure laboratory solutions do not apply.

Even assuming pure ocean water where most photosynthetic oxygen is produced, the effect of more intense short UV would be to depress the photosynthetic layer by some centimeters, possibly meters, but not to a considerable depth compared to the average depth of the oceans.

On land, a thin layer of dust will protect most biota which do not have a protective integument.

It is the writer's present opinion, therefore, that the instability of the present atmosphere oxygen concentration based primarily on the UV effect is open to some question.

3. Carbon Dioxide

Evidence that carbon dioxide is increasing in the atmosphere has been presented by various investigators, particularly Plass [31, 32, 33], Keeling [22], and Pales and Keeling [29]. According to Pales and Keeling the concentration of this gas has been increasing steadily since 1958. However, the rate has been decreasing, even though the rate of fossil fuel utilization has been increasing [22, 29].

In 1965, 1.7×10^{10} tons of CO_2 were injected into the atmosphere. From C^{14} data, the atmospheric residence of CO_2 was estimated at five years [3, 22]. A substantial part of atmospheric CO_2 becomes dissolved in ocean waters in the five-year period.

According to Kuentzel [26], the rate of photosynthesis by algae is greatly increased in the presence of additional carbon dioxide. At least with fresh water algae, it appears that the concentration of CO_2 is the rate limiting nutrient rather than phosphorus or nitrogen. Kuentzel points out that algae require very little phosphorus in their aqueous medium, of the order of 0.01 mg/L (ppm). Kuentzel attributes the presence of massive algae blooms in eutrophic lakes to additional CO_2 produced by bacteria which ferment the organic matter introduced by pollution sources such as sewage. Although nitrogen and phosphorus are important nutrients to the photosynthetic algae, they are usually present in excess, whereas carbon dioxide becomes rapidly depleted in the presence of active photosynthesis and algae growth becomes self-limiting if the main source of CO_2 is from the atmosphere.

It is interesting to speculate whether photosynthesis in the oceans by photoplankton might be increased as the result of human activity by introduction of additional carbon dioxide, either directly from the atmosphere, or indirectly from organic matter and sewage from land runoff, outfalls and barged wastes. The continental coasts of the United States, excluding Alaska, receive 10^{12} gallons of water per day from land runoff. Substantial organic matter is introduced to coastal waters by this route. The major sources of the organic matter are soil humus, farm wastes, and human sewage. The latter has been increasing with the population which now exceeds 200 million. Since livestock is raised in proportion to the population, farm wastes are also increasing. No figures are available on the average organic content of waters reaching the coasts. Raw sewage has an oxygen demand of 0.0011 pounds O_2 per gallon. Farm wastes should have a similar oxygen demand and, volume-wise, should exceed human wastes by several times. Moreover, farm wastes are not usually treated in sewage disposal plants. It can be safely assumed, therefore, that surface water draining to the continental coasts contains a substantial amount of organic matter contributed directly or indirectly by human activity.

It seems quite evident that the ocean waters surrounding the continental United States are receiving more nutrients in the form of organic matter, factory-made chemical fertilizers, and carbon dioxide from the atmosphere at a rate exceeding that a millenium ago. If a similar process is occurring on a global basis, a reasonable assumption, considering world population growth, is that photosynthesis may be increasing. If this actually is the case, then one might expect an increase in the rate of oxygen replenishment to the atmosphere from the ocean surfaces. Unless excess oxygen production from ocean waters is exactly compensated by combustion of fossil fuels and oxidation of recent organic matter on the continents, there should be an increase in the relative concentration of oxygen in the atmosphere. The residence time of oxygen in today's atmosphere should be less than that of a thousand years ago, in any case.

We are now witnessing man-made eutrophication of large lakes; e.g., Lake Erie. Can Kuentzel's interesting observation on stimulation of photosynthesis from fresh water algae by additional CO_2 be extended to ocean photosynthesis? Compared with relatively unpolluted fresh water in equilibrium with the atmosphere, sea water contains dissolved CO_2 (as CO_2, H_2CO_3, HCO_3^-, and $CO_3^=$) in much higher concentrations (see Table I). The introduction of additional CO_2 to sea water might not have the

TABLE I

Relative concentration of gases in the atmosphere and in sea water

Gas	*A* Atmosphere[a]	*B* Sat. sea water[b] ($S = 34.3$ ‰) ($T = 10$°C)	$\frac{B \times 10^{-1}}{A}$
	Mol. fraction %	ml/L ‰	
Nitrogen (N_2)	78.09	11.56	0.015
Oxygen (O_2)	20.95	6.44	0.031
Argon (A)	0.93	0.3	0.032
Carbon dioxide (CO_2)	0.03	34–56	113–188
Neon (Ne)	2×10^{-3} }	1.7×10^{-4}	
Helium (He)	5×10^{-4} }		
Krypton (Kr)	1×10^{-4}		
Hydrogen (H_2)	5×10^{-5}	3×10^{-3}	0.015
Xenon (Xe)	8×10^{-6}		
Ozone (O_3)	1×10^{-6}		
Radon (Rn)	6×10^{-8}		
H_2S	?	0–22	
CH_4	?	0–30	

[a] Values obtained from *Handbook of Chemistry and Physics*, 47th Edition, represent sea-level atmospheric composition for a dry atmosphere and do not necessarily indicate exact condition of atmosphere.
[b] Values of major constituents obtained from *The Oceans*. High values for H_2S and CH_4 found only in waters over anoxic basins.

same stimulating effect as described by Kuentzel's fresh water system. On the other hand, the introduction of any agent to sea water that might cause a shift in pH to the acid side might cause a dramatic increase in available CO_2 with consequent increase in the photosynthesis rate. The reserve of CO_2 in sea water as carbonate is not directly available to photoplankton.

Moreover, the total carbon dioxide concentration in all molecular, ionic, and mineral forms is decreased in sea water at higher pH values. There is usually an alkaline shift associated with active photosynthesis resulting in a tendency for carbonate to precipitate out of solution. Because of the unique equilibrium of the carbon dioxide system in sea water, the oceans' major buffering mechanism, the availability of this essential nutrient becomes self-limiting to photosynthesis. Keeling [22] assumes the rate of CO_2 plant uptake on land to be proportional to the CO_2 partial pressure of the atmosphere. This follows Kuentzel's observations of fresh water algae. Whether additional CO_2 reaching the oceans will increase the rate of photosynthesis remains to be verified.

As yet there appears no firm evidence of change in relative concentrations of oxygen or nitrogen in the lower atmosphere, although local transient and diurnal changes for oxygen have been noted by some observers [10, 20]. Up until the present time, the relative concentration of oxygen in the atmosphere of approximately 21% appears to be regulated largely by the incident annual solar light reaching the earth, rather than other factors such as local concentration of nutrients. Variations in primary productivity of the oceans show a wide range in regional and seasonal differences (see Table II). Broadly speaking, the photosynthetic activity, with consequent oxygen production, ranges from 5 to 5000 mg carbon per m^2 per day. Thus, oxygen pro-

TABLE II

Oceanic productivity measurements, based on C^{14} fixation (Original sources compiled in Strickland (1960), Ryther (1963), except Equatorial Atlantic from IGY cruise data)*

Region	Time of Year	Primary productivity ($mg\ C/m^2/day$)
Atlantic Ocean areas		
Arctic		
Ice Island T3	Midsummer	0–24
Station Alpha	Late summer	0–6
Western Barents Sea, near Bear Island		
Arctic water	May	1300
Atlantic water	May	275
Norwegian Sea, near Spitzbergen		
North Atlantic water	June	2400
Arctic water	June	400–600
Northern North Atlantic		
Faroe-Iceland Ridge	Summer	650–2700
Near Iceland	Summer	530–1300
South of Greenland	Summer	550
Irminger Sea	Summer	150–250
North Sea		
Annual range	Yearly	100–1500
Northeast Coast of England	May	220
	October	110
	February	5
Danish Coastal waters	March bloom	300
	August (max)	700
	December (min)	10
Eastern North Atlantic		
15 miles off Oporto	September	100
200 miles off Oporto	September	150

Table II (continued)

Region	Time of Year	Primary productivity (mg C/m²/day)
Mediterranean Sea		
2 miles off French Coast	Midsummer	30–40
Western North Atlantic		
Continental shelf and inshore		
Spring flowering	Spring	1930
Mean rate		560
Weighted annual mean		330
Sargasso Sea	April bloom	890
	Summer	100–200
Caribbean Sea (10–20°N)		100–200
Equatorial Atlantic		
West Basin		
Near 20°N	February	60–160
At 8°15′N	April–May	230–300
At 8°15′S	March	70–280
At 15°45′S	April	20–130
East Basin		
Near 20°N	February	190–780
At 8°15′N	April–May	180–1480
At 8°15′S	March	100–370
At 15°45′S	April	90–420
Southwest Atlantic		
Walvis Bay and Benguela Current	December	500–4000
Pacific Ocean areas		
Equatorial Pacific		
Near coast of Ecuador	Autumn	500–1000
9°N, 90°W (Costa Rica dome)	November	410–800
11°N, 115°W	Autumn	10
Range, 30°N–30°S	March	100–250
Northwest Pacific		
Sea of Japan		
Kuroshio Current	Summer	50–100
Oyashio System	Summer	250–500
Sea of Okhotsk, North of Japan		
Range of values	May	6–5100
Mean value	May	2000
Indian Ocean areas		
Equatorial Indian Ocean		200–250

* From *Encyclopedia of Oceanography* (ed. by R. W. Fairbridge), 1966 (10).

duction in the oceans is unevenly distributed and varies as much as a thousand times. Since this pattern is constantly changing, it is remarkable that the concentration of oxygen in the atmosphere remains constant. Johnson [20] in this volume estimates the time constant for O_2 in the atmosphere as produced by photosynthesis to be

10^4 years, which may account for its present uniform concentration.* Whether this stability will be affected by man-made activity is still an open question.

Recent Weather Bureau reports (see also Bryson and Wendland in this volume, p. 130) state that there is a worldwide increase in particulate matter in the atmosphere which filters out some solar radiation. This suspended material is believed to be largely smoke and dust particles from burning grass lands and denuded soil from man-made activities. Evidence for the rapid buildup of atmospheric particulate matter is found in recent deposits in glaciers and high altitude snow deposits. More recently, exhausts from jet engines, factories, and motor vehicles have added to the atmosphere's particulate burden. The impact of this man-made solar filter may cause a reduction in the earth's temperature – a reverse of the 'greenhouse effect' [3]. Whether the earth is cooling off or warming up is another matter open to question. Mitchell [27] discusses this point in this volume. Whether this increase is a temporary perturbation or a more prolonged effect resulting from man-made activity is not yet established. Solar flares and storms, for example, would be an obvious cause of a temperature perturbation which is not man-made.

4. Nitrogen

The situation with respect to nitrogen is even less clear than that of oxygen. Although the writer does not feel that an upset in the nitrogen cycle is as imminently serious as with oxygen, the two are interrelated along with, also, carbon and hydrogen. Nitrogen differs from the other elements mentioned, notably in its distribution. Most all of the earth's nitrogen is in the atmosphere, with a smaller precentage tied up in the biosphere (including the hydrosphere) and some small deposits of biologic origin, such as saltpeter in Chile and oxidized organic minerals from guano deposits in a few isolated areas. Although volcanic gases show N_2, O_2, H_2S, SO_2, CO, CO_2, CH_4, HF, HCl, etc., the N_2 and O_2 are believed to be from either atmospheric contamination or from interaction with sea water. The analyses of unweathered sedimentary rock by Friedman, show very little nitrogen present [13]. The concentration of nitrogen

TABLE III

Percent concentration of oxygen and nitrogen in igneous rock

Oxygen	46.42 %
Nitrogen	0.00463 %

* Uniform concentration refers to the lower atmosphere. The situation is apparently different at higher altitudes. Norton and Warnock [28] report a seasonal variation of atmospheric molecular oxygen in the 100–200 km region as 50 ± 20 % lower in winter than in summer between latitudes 45 and 65°.

in igneous rock, moreover, is 10000 times less than that of oxygen [36]*. In brief, the earth's crust cannot provide a substantial reservoir of nitrogen as is the case with oxygen and carbon dioxide. This also applies to the hydrosphere where most of the dissolved nitrogen present originates directly from atmospheric equilibria, with a small fraction (as gas, inorganic, and organic compounds) indirectly via the biosphere. In contrast, the oceans provide a vast reservoir of oxygen from water itself since O^{18} studies show that O_2 liberated in photosynthesis originates almost entirely from H_2O and not CO_2 [16].

Although nitrogen in its various molecular and ionic forms has a potential of eight electron steps, the most common forms found in nature are nitrogen gas, oxides of nitrogen gas, organic nitrogen, ammonia, nitrite and nitrate. Microorganisms play an important role in the nitrogen cycle, affecting reversibly all natural states in redox reactions.

A simplified version of the nitrogen cycle in the biosphere as presented in many texts (see, e.g., p. 571 of [30]) is as follows:

$$\rightarrow N_2 \rightarrow RNH_2 \rightleftharpoons NH_3 \rightleftharpoons NO_2^- \rightleftharpoons NO_3^- \longrightarrow$$

where R represents an organic complex such as in amino acids and protein. Those microorganisms that liberate free nitrogen from nitrite and nitrate are called denitrifiers. The literature is not too clear as to whether denitrification is the principal process for liberation of free nitrogen from the biosphere. This gas is observed to emanate from organically-rich, anaerobic, water-saturated soils and sediments where little or no nitrite or nitrate should exist [12, 34, 37, 39]. This point is mentioned because pollution of estuaries with organic-rich wastes covered by silt could create conditions speeding up release of nitrogen gas to the atmosphere with a subsequent reduction in the retention time of nutrient nitrogen available for desirable biological activity.

A few species of microorganisms fix** nitrogen gas while many species are involved in liberating free nitrogen from organic and inorganic materials. Since nitrogen fixation requires high free energy (endergonic), whereas nitrogen liberation yields energy (exergonic), a considerable conservation of energy is realized in the biosphere

* To date, no one has obtained a positively identified uncontaminated sample of the mantle, which constitutes 67.2 % of the earth's volume [21]. Igneous rock, representative of the earth's crust, accounts for less than 1 % of the earth's volume. Some traces of nitrogen have been found in diamonds which are believed to have originated in the mantle where required high temperatures and pressures prevail for their formation [9, 21]. The mantle would seem a logical reservoir or source of nitrogen if this element is of terrestrial origin. More than 90 % of the mantle is believed to consist of compounds of four elements – magnesium, iron, silicon and oxygen [21].

** Nitrogen fixation is defined as the process whereby free nitrogen combines chemically with other elements. This process occurs on land and is encouraged in organic farming. Although nitrogen fixation by microorganisms and phytoplankton of the oceans is known to exist, the total contribution to the oceans' nitrogen budget is not clearly understood [21, 38].

if the nitrogen liberation step can be slowed down. To accomplish this, it is necessary to understand more thoroughly the precise environmental conditions which tend to keep soil or sediment nitrogen in a fixed form. Commoner [6, 7] was perhaps alluding to this when he deplored the widespread practice of using inorganic nitrogen fertilizers in place of organic forms.

Unless plants can quickly absorb the inorganic nitrogen, the excess in the soil is quickly washed into the nearest drainage system where it is not only irretrievably lost to the farmer, but where it creates local problems of eutrophication. Where organic or ammonia nitrogen is used as fertilizer, the soil tends to retain these compounds [39]. The use of manure and the practice of plowing under nitrogen-fixing leguminous plants provides not only a desirable source of nitrogen, but also a humus-type soil which acts as a blotter to retain nitrogen compounds, desirable moisture content, and other essential nutrients. Commoner [6, 7] stresses the importance of humus soils in the N cycle. Also, such a composted soil is better aerated, which should retard excessive nitrogen liberation as free gas.

In contrast, the incentives for quick cash crops have encouraged the use of inorganic fertilizers with little or no composting. In fact, some farm practices during harvest time remove the entire cash crop, roots, stems, and leaves from the ground leaving the soil bare, subject to erosion by water and wind. As to the latter, the dust bowls of the Earth apparently are on the increase, as mentioned above [3].

The nitrogen content (hence the protein content) of plants varies considerably according to the species. In the corn plant, protein approximates 10% of the entire plant [42]. Under controlled laboratory conditions, the protein content of algae (chlorella) can be made to vary from 7% to as high as 88% depending upon the nitrogen concentration used as an essential nutrient [4].

These laboratory findings suggest that natural environmental conditions affecting the nitrogen balance may have an important role in determining the quality and quantity of protein in other plants, on land and in the sea. Therefore, in considering the importance of the nitrogen cycle in food production, proper control of the environment seems highly desirable. This can readily be accomplished on land. Control is still possible in the estuaries. In the oceans, however, at the present time, about all that can be done is to observe for changes and continue research of the nitrogen cycle in this environment. As with oxygen and carbon dioxide, studies of equilibria trends and exchange processes at the sea surface should be stressed to predict possible atmospheric changes.

The origin and fate of nitrogen on earth is interesting and puzzling. As to its origin, aside from what has been discussed above, it is assumed by Rubey, Urey, and others that the present nitrogen is of terrestrial origin.

The fate of nitrogen, of course, is a question of some importance as discussed by Cole [5]. Although Cole does not elaborate on how nitrogen can be lost to earth, this deserves consideration. There seem only two possibilities; loss to outer space and nuclear transformation [16]. Both appear negligible, hence the persistence of nitrogen on earth seems assured in the foreseeable future.

5. Summary and Conclusions

The previous sections have considered some large-scale time and space aspects of the impact of modern technology and population growth on the biosphere of the oceans and continents and possible consequences to the quality of the atmosphere. It seems obvious that man can shift natural equilibrium forces that make for a healthy balance in nature essential for life, in particular the composition of the atmosphere and the food potential of land and sea.

Thus far, a major disruption in the oceans and atmosphere is not apparent. We may, however, be looking at subtle changes in the environment, such as the accumulation of lead and other toxic chemicals, an increase in carbon dioxide and organic carbon compounds, atmospheric dust, and acid rainfall which may have far-reaching consequences in years to come. The equilibrium existing between physical, chemical, geological and biological forces is a subject for serious study on a global basis.

"All things by almighty power
Near and far
Hiddenly connected are
That thou canst not pick a flower
Without disturbing of a star."

Anon

References

[1] Berkner, L. V. and Marshall, L. C.: 1965, 'On the Origin and Rise of Oxygen Concentration in the Earth's Atmosphere', *J. Atmos. Physics* **22**, 225–261.
[2] Berkner, L. V. and Marshall, L. C.: 1966, 'Potential Degradation of Oxygen in the Earth's Atmosphere', Memo for file.
[3] Bryson, R. A. and Wendland, W. M.: 1970, 'Climatic Effects of Atmospheric Pollution', Symposium on Global Effects of Environmental Pollution, AAAS National Meeting, 1968, this volume, p. 130.
[4] Burlew, J. S. (ed.): 1961, 'Algal Culture – From Laboratory to Pilot Plant', Carnegie Institution of Washington Publication #6000, Washington, D.C.
[5] Cole, LaMont C.: 1967, 'Can the World Be Saved', Paper presented at the 134th meeting of the American Association for the Advancement of Science.
[6] Commoner, B.: 1967, 'The Balance of Nature', Address to the Graduate School, U.S. Department of Agriculture, Washington, D.C.
[7] Commoner, B.: 1970, 'Threats to the Integrity of the Nitrogen Cycle: Nitrogen Compounds in Soil, Water, Atmosphere and Precipitation', Symposium on Global Effects of Environmental Pollution, AAAS National Meeting, 1968, this volume, p. 70.
[8] Conference on Physical and Chemical Properties of Sea Water, Easton, Maryland, September 4–5, 1958; Washington National Academy of Sciences, National Resource Council, 1959.
[9] De Carlie, P. S.: Stanford Research Institute, personal communication.
[10] *Encyclopedia of Oceanography. Encyclopedia of Earth Sciences Series*, Vol. 1 (ed. by R. W. Fairbridge), Reinhold Publishing Company, New York, 1966.
[11] *Equilibrium Concepts in Natural Water Systems. Advances in Chemistry Series* **67**. A Symposium Sponsored by Division of Water, Air and Waste Chemistry, 151st Meeting, American Chemical Society, Pittsburgh, Pennsylvania, March 23–24, 1966.

[12] Felbeck, G. T.: 1966, 'Normal Alkanes in Much Soil Organic Matter Hydrogenolysis Products', *Trans. Comm. II and IV, Int. Soc. Soil Sci.*, Aberdeen.
[13] Friedman, I.: U.S. Geological Survey, personal communication.
[14] Gambel, A. W. and Fisher, D. W.: 1966, 'Chemical Composition of Rainfall in Eastern North Carolina and Southeastern Virginia', Geological Survey Water Supply Paper #1535-K.
[15] Gambell, A. W. and Fisher, D. W.: 1966, 'Chemistry of Atmospheric Precipitation', U.S. Geological Summary Report ACA-17-F.
[16] Glasstone, S.: 1958, *Sourcebook on Atomic Energy*, Van Nostrand, New York, 2nd Edition.
[17] Goldberg, E. D.: 1970, 'The Chemical Invasion of the Oceans by Man', Symposium on Global Effects of Environmental Pollution, AAAS National Meeting, 1968, this volume, p. 178.
[18] *Handbook of Chemistry and Physics*, 47th edition, The Chemical Rubber Co., 1967.
[19] Hires, R. I., Stroup, E. D., and Seitz, R. C.: 1963, 'Atlas of the Distribution of Dissolved Oxygen and pH in Chesapeake Bay 1949–1961', Chesapeake Bay Inst. Graphical Summary Report No. 3.
[20] Johnson, F. S.: 1970, 'The Oxygen and Carbon Dioxide Balance in the Earth's Atmosphere', Symposium on Global Effects of Environmental Pollution, AAAS National Meeting, 1968, this volume, p. 4.
[21] Josephs, M. J. and Sanders, H. J.: 1967, 'Chemistry and the Environment', Am. Chem. Soc. Pub., Washington, D.C.
[22] Keeling, C. D.: 1968, 'Carbon Dioxide from Fossil Fuel – Its Effect on the Natural Carbon Cycle and on the Global Climate', 49th Meeting, American Geophysical Union, Washington, D.C.
[23] Keeling, C. D.: 1968, 'Carbon Dioxide in Surface Ocean Waters: Global Distribution', *J. Geophys. Res.* **73**, 4543–4553.
[24] Keeling, C. D., Harris, T. B., and Wilkins, E. M.: 1968, 'Concentration of Atmospheric Carbon Dioxide at 500 and 700 Millibars', *J. Geophys. Res.* **73**, 4511–4528.
[25] Keeling, C. D. and Waterman, L. S.: 1968, 'Carbon Dioxide in Surface Ocean Waters', *J. Geophys. Res.* **73**, 4529–4541.
[26] Kuentzel, L. E.: 1969, 'Bacteria, Carbon Dioxide and Algal Blooms', *Proceedings of the 24th Annual Purdue Industrial Waste Conference*, Purdue University, Lafayette, Indiana (in press).
[27] Mitchell, J. M., Jr.: 1970, 'A Preliminary Evaluation of Atmospheric Pollution as a Cause of Global Temperature Fluctuations', Symposium on Global Effects of Environmental Pollution, AAAS National Meeting, 1968, this volume, p. 139.
[28] Norton, R. B. and Warnock, J. M.: 1968, 'Seasonal Variation of Molecular Oxygen Near 100 Kilometers', *J. Geophys. Res.* **73**, 5798–5800.
[29] Pales, J. C. and Keeling, C. D.: 1965, 'The Concentration of Atmospheric Carbon Dioxide in Hawaii', *J. Geophys. Res.* **70**, 6053–6076.
[30] Pelczar, J. J., Jr. and Reid, R. D.: 1965, *Microbiology*, 2nd edition, McGraw-Hill, New York.
[31] Plass, G. N.: 1956, 'The Carbon Dioxide Theory of Climatic Changes', *Tellus* **8**, 140.
[32] Plass, G. N.: 1957, *Proc. Conf. on Research in Climatology*, Scripps Inst. Oceanog.
[33] Plass, G. N.: 1956, 'Carbon Dioxide and the Climate', *Amer. Scientist* **44**, 302.
[34] Preul, H. C. and Schsoepfer, G. J.: 1968, 'Travel of Nitrogen in Soils', *J. Water Pollution Control Federation* **40**, 30–48.
[35] Rabinowitch, E. I.: 1945, *Photosynthesis and Related Processes*, Interscience, New York.
[36] Rankama, K. and Sahama, Th. G.: 1950, *Geochemistry*, University of Chicago Press, Chicago.
[37] Sisler, F. D.: 1966, 'Role of Earth Potentials in Organic Geochemical Processes Concerned with Petroleum', *Abh. Dent. Akad. Wiss. Berlin. Kl. Chem. Geol. Biol.* 199–204.
[38] Sisler, F. D. and ZoBell, C. E.: 1951, 'Nitrogen Fixation by Sulfate-reducing Bacteria Indicated by Nitrogen/Argon Ratios', *Science* **113**, 511–512.
[39] Stewart, W. D. P.: 1967, 'Nitrogen-Fixing Plants', *Science* **158**, 1426–1432.
[40] Sverdrup, H. U., Johnson, M. W. and Fleming, R. H.: 1942, *The Oceans*, Prentice-Hall, New York.
[41] Tatsumoto, M. and Patterson, C. C.: 1963, 'The Concentration of Common Lead in Sea Water' in *Earth Science and Meteorities* (ed. by J. Geiss and E. D. Goldberg), North-Holland Publishing Co., Amsterdam.
[42] Watt, B. K. and Merrill, A.: 1963, 'Composition of Foods', *Agri. Handbook*, No. 8, U.S. Department of Agriculture.

For Further Reading

1. Effects of population growth and technology on pollution of the sea:
   Panel Reports of the Commission on Marine Science, Engineering and Resources, Vol. 1: *Science and Environment*, U.S. Government Printing Office, Washington, D.C. 1969.
2. Significance of photosynthesis in oxygen and carbon dioxide equilibria:
   E. I. Rabinowitch, *Photosynthesis*, Vol. 1 and 2, Interscience Publishers, New York, 1956, 2088 pp.
3. Pollution from large-scale use of nitrogen fertilizers and other agricultural chemicals:
   U.S. Dept. of Agriculture, *Wastes in Relation to Agriculture and Forestry*, Pub. No. 1065, U.S. Government Printing Office, Washington, D.C., 1968.
4. Chemical and physical interactions between the earth, oceans and atmosphere:
   Special report from Chemical and Engineering News, American Chemical Society, *Chemistry and Environment*, A.C.S. Publications 1967.

THE DEPENDENCE OF ATMOSPHERIC TEMPERATURE ON THE CONCENTRATION OF CARBON DIOXIDE

SYUKURO MANABE

Geophysical Fluid Dynamics Laboratory/ESSA, Princeton, N.J., U.S.A.

Abstract. Numerical computations using a radiative, convective equilibrium model of the atmosphere predict an increase of 0.8 °C in temperature of the earth's surface by the end of this century, based on the anticipated increase in CO_2.

1. Introduction

It seems to be certain that the concentration of carbon dioxide in the atmosphere is indeed increasing with time. I shall discuss how the world temperature may be affected by such an increase.

As you know, carbon dioxide is nearly transparent to visible light but it is a strong absorber of infrared radiation particularly in the wavelength from 12 to 18 microns; consequently, an increase of atmospheric carbon dioxide could act much like a glass in a greenhouse to raise the temperature of the lower atmosphere.

The dependence of world temperature upon the concentration of carbon dioxide has been evaluated by various authors, e.g., Plass (1956), Kaplan (1960), Kondratiev and Niilisk (1960), and Möller (1963). The latter's conclusion is quite different from those of preceding authors. One shortcoming of these studies is that their estimates were obtained from the computation of the heat balance of the earth's surface instead of the atmosphere as a whole. Here, I would like to discuss the possibility of estimating this dependence by using the mathematical model of the whole atmosphere.

2. Calculation Procedure

Two of the most fundamental processes controlling the thermal structure of the atmosphere are radiative transfer, and moist or dry convection. Recently, we have been successful in obtaining the state of radiative, convective equilibrium of the atmosphere from the numerical integration of the model with both of these two processes (Manabe and Strickler, 1964; Manabe and Wetherald, 1967). By comparing the states of equilibrium, which were obtained for various CO_2 concentrations, it was possible to estimate the dependence of atmospheric temperature upon CO_2 concentration.

Although the major constituents of the earth's atmosphere are nitrogen and oxygen, they hardly absorb the atmospheric radiation. However, minor constituents such as water vapor, carbon dioxide and ozone, have strong absorption bands and affect the field of both solar and terrestial radiation. In our computation, the radiative effects of these gaseous absorbers as well as that of clouds are calculated by using the equation of radiative transfer.

The state of radiative convective equilibrium was approached asymptotically by the numerical time integration of the model starting from the initial condition of an isothermal atmosphere. In order to simulate the macroscopic behavior of moist convection, we introduced a very simple concept of a so-called 'convective adjustment'. Whenever the vertical temperature gradient exceeds the neutral gradient for moist convection, it was assumed that the neutral lapse rate* is restored instantaneously by the effect of the free moist convection.

3. Results

Figure 1 shows the approach towards the state of equilibrium. Towards the end of this time integration, the magnitude of the net downward solar radiation is almost exactly equal to that of net upward terrestial radiation at the top of the atmosphere, i.e., the atmosphere is in complete thermal equilibrium as a whole. In Figure 2, the state of radiative, convective equilibrium for the hemispheric mean insolation is compared with the U.S. standard atmosphere. The agreement between the two distributions is excellent.

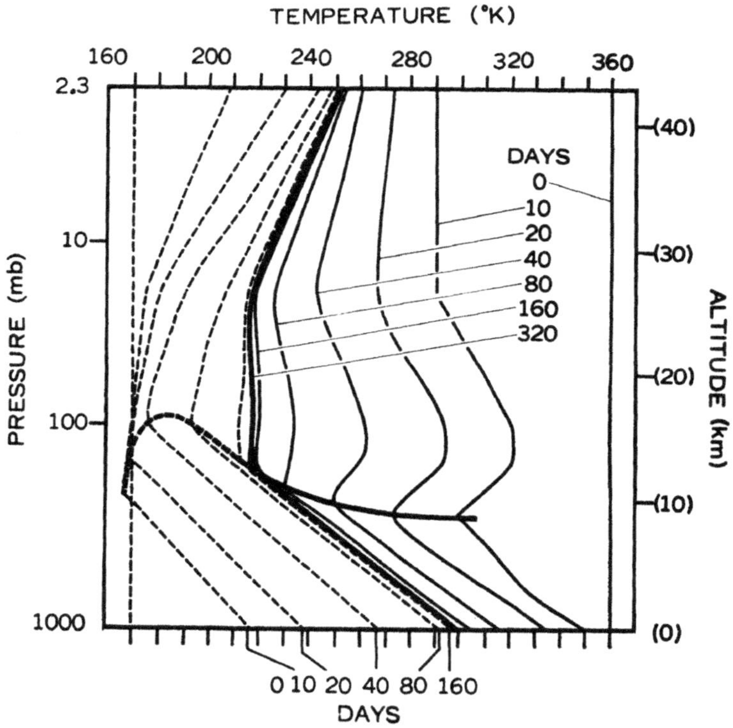

Fig. 1. Approach towards the state of radiative, convective equilibrium. The solid and dashed lines show the approach from a warm and cold isothermal atmosphere. (By Manabe and Strickler, 1964).

* The neutral lapse rate is assumed to be 6.5°C/km for this study.

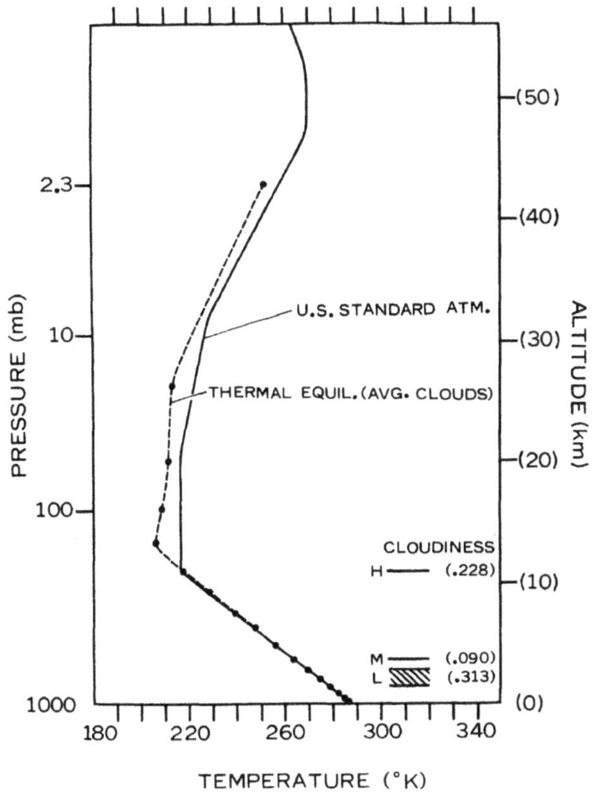

Fig. 2. Dashed line shows the radiative, convective equilibrium of the atmosphere with cloudiness indicated in the right hand side of the figure. The solid line shows the U.S. Standard Atmosphere. (By Manabe and Strickler, 1964).

Encouraged by this agreement, we decided to evaluate the dependence of the equilibrium temperature upon the concentration of carbon dioxide by using this model. In the equilibrium computation described so far, the state of radiative equilibrium has been obtained for the given distribution of absolute humidity. It is well known, however, that the warmer the atmosphere is, the more moisture it usually contains. As Möller (1961) pointed out, the atmosphere tends to preserve the general level of relative humidity rather than that of absolute humidity through the process of condensation and evaporation. Owing to the dependence of the so-called greenhouse effect upon the concentration of water vapor, the equilibrium temperature of the atmosphere with a given distribution of relative humidity is almost twice as sensitive to the change of CO_2 concentration as that of the atmosphere with a given distribution of absolute humidity. Table I shows the results of our computation.

This table indicates that the doubling or halving the CO_2 concentration increases or decreases the surface temperature of the atmosphere by about 2.3 °C. Suppose the concentration of CO_2 increases by about 25% from AD1900 to AD2000 as the U.N. Dept. of Social and Economic Affairs predicts, the resulting increase of surface

TABLE I

Change of equilibrium temperature of the earth's surface in °C corresponding to various changes of CO_2 content of the atmosphere (Manabe and Wetherald, 1967).

Change of CO_2 content (ppm)	Fixed absolute humidity		Fixed relative humidity	
	Average cloudiness	Clear	Average cloudiness	Clear
300→150	−1.25	−1.30	−2.28	−2.80
300→600	+1.33	+1.36	+2.36	2.92

temperature would be about 0.8 °C, which may have significant effect upon the climate of the earth's surface. Figure 3 shows how the vertical distribution of temperature depends upon the CO_2 content. It is interesting that in the stratosphere, the larger the CO_2 concentration is, the colder is the temperature.

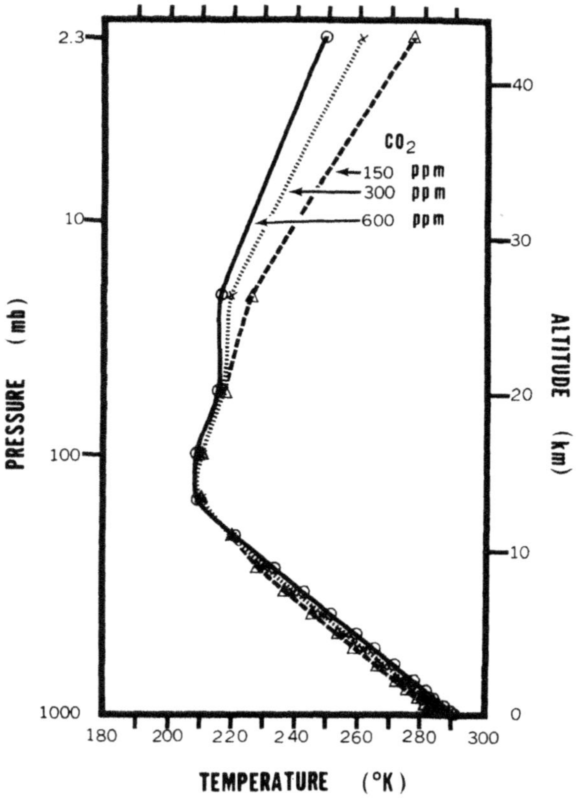

Fig. 3. Vertical distribution of temperature in radiative, convective equilibrium for various values of CO_2 content. (By Manabe and Wetherald, 1967).

4. Discussion

So far, we have discussed the dependence of equilibrium temperature upon the concentration of carbon dioxide. In order to discuss how the latitudinal distribution of temperature is affected by the change of CO_2 concentration, it is necessary to construct the three-dimensional model of the atmosphere which involves not only the effect of radiation and convection but also the dynamics of the large scale eddies and the hydrologic cycle of water and snow.* Such a model has been constructed at the Geophysical Fluid Dynamics Laboratory of ESSA. We are preparing to carry out a series of numerical experiments for various CO_2 concentrations by using this model.

In order to discuss the very long range evolution of climate it is necessary to consider the exchange of CO_2 between the ocean and the atmosphere. Recently Manabe and Bryan (1969) have attempted to construct a general circulation model of a joint ocean-atmosphere system, which could be used for such a purpose. One of the major difficulties in constructing such a model is our lack of knowledge about the distribution of the coefficient of vertical mixing by the small scale eddies in the ocean. I hope the improvement of the knowledge of the turbulent process in the ocean will enable us to study the probability of climatic instability by using such a joint model.

References

Kaplan, L. D.: 1960, 'The Influence of Carbon Dioxide Variations on the Atmospheric Heat Balance', *Tellus* **12**, 204–208.

Kondratiev, K. Y. and Niilisk, H. I.: 1960, 'On the Question of Carbon Dioxide Heat Radiation in the Atmosphere', *Geofis. Pura Appl.* **46**, 216–230.

Manabe, S., and Bryan, K.: 1969, 'Climate Calculation with a Combined Ocean-Atmosphere Model', *J. Atmos. Sci.* **26**, 786–789.

Manabe, S. and Strickler, R. F.: 1964, 'Thermal Equilibrium of the Atmosphere with a Convective Adjustment', *J. Atmos. Sci.* **21**, 361–385.

Manabe, S. and Wetherald, R. T.: 1967, 'Thermal Equilibrium of the Atmosphere with a Given Distribution of Relative Humidity', **24**, 241–259.

Möller, F.: 1963, 'On the Influence of Changes in the CO_2 Concentration in Air on the Radiation Balance of the Earth's Surface and on the Climate', *J. Geophys. Res.* **68**, 3877–3886.

Plass, G. N.: 1956, 'The Carbon Dioxide Theory of Climatic Change', *Tellus* **8**, 140–154.

United Nations Department of Economic and Social Affairs: 1956, 'World Energy Requirement in 1975 and 2000', in *Proceedings of the International Conference on the Peaceful Uses of Atomic Energy*, pp. 3–33.

General Bibliography

For the general discussion of the subject, see

Environmental Pollution Panel, President's Science Advisory Committee. 1965 'Restoring the Quality of our Environment. Appendix Y4, Atmospheric Carbon Dioxide', pp. 111–133, White House, Nov., 1965. At the end of this Appendix, one can find convenient references of this subject.

* Since snow or ice has a large reflectivity for solar radiation, the incorporation of the snow hydrology into the model may significantly increase the sensitivity of the model climate to the change in the amount of atmospheric absorbers such as CO_2.

EXCHANGE OF CO_2 BETWEEN ATMOSPHERE AND SEA WATER: POSSIBLE ENZYMATIC CONTROL OF THE RATE*

RAINER BERGER and WILLARD F. LIBBY

Institute of Geophysics, University of California, Los Angeles, Calif., U.S.A.

Abstract. Surface and sub-surface ocean water differ in exchange characteristics with atmospheric CO_2. The possibility of control by an enzyme-like carbonic anhydrase is discussed.

It has been discovered [1] that sea waters can differ markedly in their rates of equilibration with atmospheric CO_2. Surface waters on the average possess less than half the bomb C^{14} content of tropospheric air and are rising only in a matter of years [2, 3] to full equilibrium.

It has been well known for some time now [4] that in the analogous problem in mammals – the ready elimination of CO_2 from the blood for expiration in the lungs – that a special enzyme, carbonic anhydrase, is essential. So the question arises: Is there, perhaps, enzymatic control over the interchange of carbon dioxide between the atmosphere and the oceans?

Our procedure has been to take 50-gallon sea water samples in polyethylene lined barrels to China Lake in order to avoid the Los Angeles smog and to vigorously aerate them with clean desert air (some 200 liters per hour) for periods of days and then to test for the bomb C^{14} content. Figure 1 shows the course of the bomb C^{14}

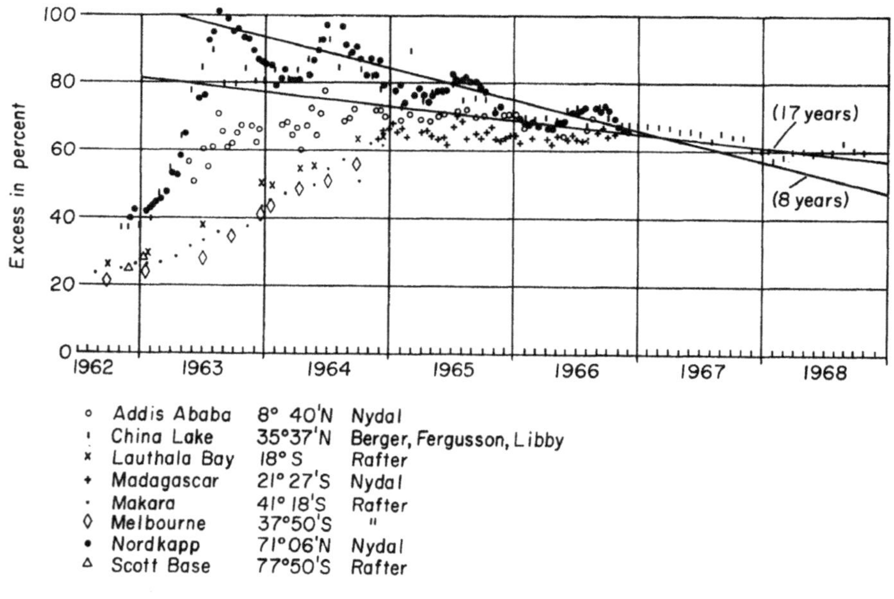

Fig. 1. Radiocarbon in atmospheric CO_2.

* Publication No. 730 of the Institute of Geophysics, University of California.

content of the surface Mojave Desert air at this location over the last seven years (analogous data from elsewhere in the world are included for comparison).

Qualitatively our results show that

(1) the exchange rate of surface waters (Santa Monica Beach) is very slow ($\sim 1/470\ \text{day}^{-1}$ unimolecular rate constant).

(2) carbonic anhydrase at 100 mg for 50 gallons increases the rate very substantially (about a factor of 20) as do 10 mg quantities.

(3) the rate of water from 200 feet depth is rapid and shows about the same characteristics as surface water to which the carbonic anhydrase has been added.

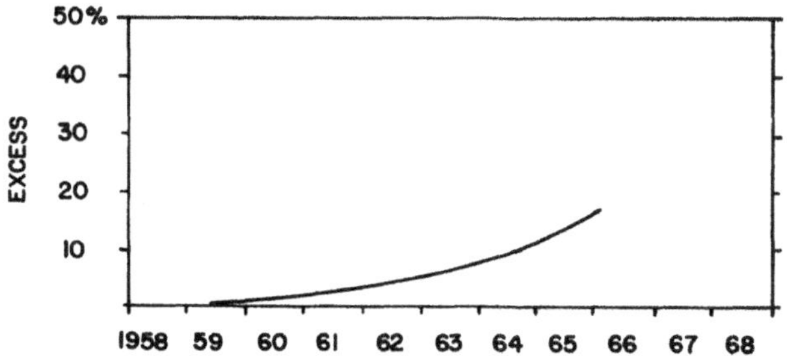

Fig. 2. Radiocarbon in ocean at 50°N.

TABLE I

CO_2 Exchange Rates for Sea Water

(50-gallon samples in polyethylene lined barrels capped, trucked to China Lake, and aerated at 200 l/hr for indicated period with or without addition of the enzyme carbonic anhydrase.)

A. Surface waters (Santa Monica Beach at foot of Sunset Boulevard)

Sample No.	Date	Initial ΔC^{14} %	Treatment	Final ΔC^{14}	Exchange Time (Days, cf. Table II)
1	12/22/65	14.9	17 days	17.3	460
2	1/8/66	15.1	68 days	22.7	480
3	4/12/66	15.4	100 mg CA + 14 days	47	19
4	5/12/66	14.0	100 mg CA + 3 days	29.6	10

B. Waters from 200 feet depth

Sample No.	Date	Initial ΔC^{14} %	Treatment	Final ΔC^{14}	Exchange Time (Days, cf. Table II)
5	3/9/67 200′ well at Naval Pt. Mugu	4.7	7.6 days	23.9	14
6	5/26/67 200′ Submarine USS Baya 33°20′N (Off Catalina Is.) 118°17′W	6.7	3.0 days	34.2	5
7	7/25/68 200′ Submarine USS Baya 31°40′N (Off Catalina Is. in 2200 fathoms total depth) 120°20′W	20.7	2.7 days	37.1	5

The general level of bomb C^{14} in surface sea waters as reported by Münnich and Roether [3] in 1967 are given in Figure 2. Comparison of Figures 1 and 2 shows that after several years the sea still lags substantially behind tropospheric air.

In order to test our theory of a biochemically controlled rate analogous to that found in mammalian systems we have studied the waters off the Southern California Coast both at the surface (Samples 1–4) and at 200 feet depth (Samples 5–7). The data are assembled in Table I.

TABLE II

Exchange rate calculations

$$C^{14}O_{2a} + CO_{2s} = CO_{2a} + C^{14}O_{2s}$$

$$\frac{[C^{14}O_{2a}]}{[CO_{2a}]} = \gamma_a$$

$$\frac{[C^{14}O_{2s}]}{[CO_{2s}]} = \gamma_t \qquad \text{Eq. 1}$$

$$[C^{14}O_{2s}] = \gamma_0 + (\gamma_a - \gamma_0)(1 - e^{-t/\tau})$$

$$[CO_{2s}] = \gamma_0 e^{-t/\tau} + \gamma_a (1 - e^{-t/\tau})$$

γ_t, γ_0 are measured, $(\gamma_a - \gamma_0)$ calculated from Figure 1 and τ is calculated from Equation (1).

$$1 - \frac{(\gamma_t - \gamma_0)}{(\gamma_a - \gamma_0)} = \exp(-t/\tau)$$

$$\tau = \frac{t}{-\ln\left(1 - \frac{(\gamma_t - \gamma_0)}{(\gamma_a - \gamma_0)}\right)}$$

Sample	γ_0	γ_a	γ_t	t(days)	$\gamma_t - \gamma_0$	$\gamma_a - \gamma_0$	ln	τ(days)
1	14.9	80	17.3	17	2.4	65.1	– 037	460
2	15.1	75	22.7	68	7.6	60.0	–.14	480
3	15.4	75	47.0	14	31.6	60.0	–.75	19
4	14.0	75	29.6	3	15.6	61.0	–.31	10
5	–4.7	65	23.9	7.6	28.6	70.0	–.53	14
6	6.7	65	34.2	3.0	27.5	58.0	–.63	5
7	20.7	63	37.1	2.7	17.1	42.0	–.53	5

It is clear that Santa Monica beach water lacks some quality which Pt. Mugu and Catalina waters from 200 feet depth possess. This quality is matched by 100 mgms of carbonic anhydrase added to 50 gallons.

Since only limited localities have been studied no very general conclusions can be drawn but the results suggest that

(1) the quality may be due to an enzyme-like carbonic anhydrase which might well be isolatable and may be derived from sea life known to produce it [5–8].

(2) there may be extensive areas of the sea devoid of the quality imparted by carbonic anhydrase and therefore slow to dissolve CO_2.

(3) that this quality may not long survive contact with air.

The rates measured here for Santa Monica beach surface waters agree well enough with the work of others [3] for the seas as a whole so we can have some confidence in the tentative conclusions. However, additional work is necessary to isolate and identify the enzyme and to measure its oxidative stability and to assay the waters of the seas for it.

Acknowledgements

We thank Captains V. H. L. Duckett and W. A. White, officers and men of USS 'Baya', G. Plain, USNWC China Lake for assistance and L. Provasoli, Haskins Laboratories, New York and J. D. Strickland, Scripps Institute of Oceanography for advice.

This research was supported in part by the National Science Foundation G-628.

References

[1] Berger, R. and Libby, W. F.: 1968, UCLA Radiocarbon Dates VII, *Radiocarbon* **10**, 149; 1969, *ibid.* **11**, No. 1, 194.
[2] Nydal, R.: 1968, *J. Geophys. Res.* **73**, 3617.
[3] Münnich, K. O. and Roether, W.: 1967, *Radioactive Dating and Methods of Low Level Counting*, International Atomic Energy Agency, Vienna 1967, p. 93–104.
[4] White, A., Handler, P., and Smith, E. L.: 1964, *Principle of Biochemistry*, McGraw-Hill, p. 663.
[5] Wilbur, K. M. and Owen, G.: 1964, in *Physiology of Mollusca* (ed. by K. M. Wilbur and C. M. Yonge), Academic Press, New York. p. 232 (for molluscs).
[6] Nicol, J. A. C.: 1968, *The Biology of Marine Animals*, Sir Isaac Pitman & Sons, London, p. 191 (for fish).
[7] Goreau, T.: 1961, *Endeavour* **20**, 32 (for corals).
[8] Isenberg, H. D., Lavine, L. S., and Weissfellner, H.: 1963, *J. Protozoology* **10**, 477 (for coccoliths).

For Further Reading

1. M. N. Hill (ed.), *The Sea*, Vols. 1, 2, and 3, Interscience Publishers, 1962.
2. C. E. Junge, *Air Chemistry and Radioactivity*, Academic Press, New York, 1963.
3. Y. Bottinga and H. Craig, 'Oxygen Isotope Fractionation Between CO_2 and Water and the Isotopic Composition of Marine Atmospheric CO_2', *Earth and Planetary Science Letters* **5** (1969), 285.
4. P. Kilho Park, 'Oceanic CO_2 System: An Evaluation of Ten Methods of Investigation', *Limnology and Oceanography* **14** (1969), 179.

THE GLOBAL BALANCE OF CARBON MONOXIDE

LOUIS S. JAFFE

Abstract. Carbon monoxide, the most abundant air pollutant found in the atmosphere, generally exceeds that of all other pollutants combined (excluding CO_2). Carbon monoxide is principally a man-made pollutant. Of the estimated 102 million tons of CO emitted in the U.S.A. during 1968 more than 90×10^6 tons were emitted from major technological sources. The motor vehicle contributed approximately 63% of the man-made CO emissions or about 58% of CO from all sources in the U.S.A. and also represents the largest single source of CO globally.

Other sources of CO include emissions from coal and fuel oil burning, industrial processes, and solid waste combustion. Some CO is also formed naturally by certain types of vegetation and marine organisms (siphonophores).

Carbon monoxide is relatively chemically inert in the troposphere; the chemical oxidation of CO in the lower atmosphere by molecular oxygen is very slow. The exact duration of CO in the lower atmosphere is not known with certainty; however, the mean residence time has been variously estimated to be between 0.1 and about 5.0 years. In the absence of scavenging processes, the estimated world-wide CO emission would be sufficient to raise the atmospheric level by 0.04 ppm per year, yet the background levels of CO in clean air do not appear to be increasing. Several potential sinks or scavenging processes are discussed. Knowledge of the precise mechanism or mechanisms of CO removal from the lower atmosphere is unsatisfactory at the present time.

1. Introduction

Carbon monoxide (CO) is the most abundant and widely distributed air pollutant found in the lower atmosphere excepting CO_2. Emissions of CO from combustion sources are second only to CO_2 in total magnitude and generally exceed that of all other pollutants combined, particularly in urban atmospheres.* Carbon monoxide is principally a man-made pollutant. Approximately 63% of technologically formed CO emissions or about 58% of CO from all sources in the U.S.A. is derived from the combustion of fossil fuels in motor vehicles. The automobile is also estimated to contribute the principal source of CO emissions globally. Smaller amounts are contributed by stationary combustion sources, industrial processes, and solid waste combustion. Some CO is also produced by natural sources.

Carbon monoxide is relatively chemically inert and is oxidized very slowly by molecular oxygen in the lower atmosphere. Although the tonnage of CO emitted to the atmosphere continues to increase with the world-wide increased consumption of fossil fuels, yet the background levels of CO in clean air do not appear to be increasing significantly. It has been postulated, therefore, that one or more scavenging processes or 'sinks' for CO exist in the atmosphere [1].

2. Sources of Carbon Monoxide

A. TECHNOLOGICAL SOURCES

Carbon monoxide is produced globally in large quantities by the incomplete com-

* Carbon dioxide (CO_2) is not normally regarded as an air pollutant *per se* in the common usage of the term but rather is the normal end product of combustion of organic fuels and substances.

bustion of carbonaceous materials used as fuels for transportation and heating; it also is generated in industrial processing and refuse burning.

More than 94 million tons of CO was emitted by these major technological sources in the U.S.A alone during 1966 with a total of 102 million tons emitted from all sources [2] (see Table I). The principal source, about 64 million tons or about 68% of the total technological sources was the combustion of fossil fuels in internal combustion engines.

TABLE I

Carbon monoxide emissions by source category in 1968 – U.S.A.

Source	CO emission (10^6 tons/yr.)			Percent of total		
A. *Technological sources*						
1. Fuel combustion in mobile sources	63.8			67.5		
Motor vehicles		59.2			62.7	
Gasoline			59.0			62.5
Diesel			0.2			0.2
Aircraft		2.4			2.5	
Vessels		0.3			0.3	
Railroads		0.1			0.1	
Non-highway use of motor fuels		1.8			1.9	
2. Fuel combustion in stationary sources	1.9			2.0		
Coal		0.8			0.8	
Fuel oil		0.1			0.1	
Natural gas		[a]			[a]	
Wood		1.0			1.0	
3. Industrial processes	11.2			11.9		
4. Solid waste combustion	7.8			8.3		
5. Miscellaneous: man-made fires, coal refuse, etc.	9.7			10.3		
Cigarette smoke		<0.01			[a]	
Technological sources (sub-total)	94.4			100		
B. *Natural sources*						
1. Forest fires	7.2					
Total from all sources	101.6					

[a] Negligible.

Fuel combustion in gasoline-powered motor vehicles, the major subcategory in this class, alone accounted for more than 59 million tons, or approximately 63% of the total CO emission from man-made sources in the U.S.A. Aircraft emissions accounted for about 2.4 million tons, or 2.5% of this total. Global CO emissions from technological sources and forest fires have recently been estimated to approximate 230 million tons a year [3]. Motor vehicle exhaust is also by far the largest single source globally. The aforementioned estimates for CO emissions in the U.S.A. and world-wide include CO estimates of technological origin and of forest fires only. They do not include any estimates of CO from other natural sources, whether geophysical or biological in origin.

Industrial processes and miscellaneous man-made combustion (including structural and agricultural fires, coal banks, detonation of explosives, and the smoking of cigarettes) are the second and third largest technological sources of CO, while solid waste combustion is the fourth largest technological source of CO in the U.S.A. [1, 6] (see Table I) as well as globally [3]. Although relatively insignificant amounts of atmospheric CO are produced by cigarettes, cigarette smoke is, nevertheless, a major source of personal air pollution in smokers [4, 5]. Carbon monoxide emission from cigarettes is estimated to be about 0.01 million tons per year based on an annual consumption of 539 billion cigarettes that emit on the average of 16.2 mg per cigarette (see Table I).

B. NATURAL SOURCES

No large natural sources of CO have heretofore been positively identified. Recently, however, a number of geophysical and biological sources of CO, the quantities of which are presently unknown but estimated to be relatively small, have been investigated and reported [1, 6].

1. *Geophysical Sources*

Some CO is reported to be produced in volcanic and marsh gases, and natural gases found in coal mines [7]. It has also been reported that some CO is formed during electrical storms [8]. Atmospheric CO has been discovered in the solar spectrum [9]. All these findings indicate that CO is a world-wide constituent of the atmosphere. Some CO is also believed to be formed in the upper atmosphere (above 70 km) by photo-dissociation of CO_2 [10].

Another natural source of CO is the result of photochemical degradation of various reactive organic compounds involved in the formation of photochemical smog [11, 12]. The amount of CO so formed is difficult to estimate, for it is dependent not only on the kinds and types of reactive organic compounds present, but also on the intensity of sunlight present.

Forest and prairie fires contribute a substantial portion of atmospheric CO emissions in the U.S.A. (see Table I), as well as globally [3]. Those formed by electrical storms are clearly a natural source.

2. *Biological Sources*

Small quantities of CO are formed by vegetation during seed germination and seedling growth of higher plants [13, 14], and by certain marine brown algae or kelps [15, 16]. Float cells of the kelp *Nereocyctis* have been found to have CO concentrations of up to 800 ppm [3]. Microorganisms have been shown to produce CO from plant flavonoids [17]. No estimates of the world-wide emissions of CO by land-based, littoral, or pelagic vegetation are available.

Carbon monoxide is also produced by colonies of marine hydrozoan jelly-fish known as siphonophores [18, 19]. These invertebrate colonies are widespread and make up a large portion of the plankton in the warmer oceans of the world. Carbon monoxide is also produced in the float cells of *Physalia physalis* (Portuguese Man-of-War), a surface dwelling siphonophore [20]. The carbon monoxide generated by kelps and the millions of siphonophores found in the oceans of the world contribute substantially to the relatively high concentrations of CO found in sea water.

Another biological source of CO is endogenous CO produced in man and animals as a by-product of heme catabolism [21].

The total CO formed and emitted by these natural sources is difficult to calculate. No estimates of CO emissions exist except for the recent oceanic data of Swinnerton *et al.* [22]. Measurements of the distribution of CO between the atmosphere and surface waters of the western Atlantic show that the waters are supersaturated with respect to the partial pressure of this gas in the atmosphere. Under these conditions the net transport must be from the sea to the atmosphere. Assuming similar production rates throughout the world oceans, these investigators indicate that the ocean may be the largest known natural source of CO, contributing 9×10^{12} g (1×10^7 tons) or roughly about 5% of the CO from technological sources [22].

3. Characteristics of Environmental CO in Community Atmospheres

A. TEMPORAL VARIATIONS

The concentration of CO in metropolitan areas varies widely with time and place and is dependent on human activity and meteorological factors [23–32]. Continuous air monitoring of CO at Continuous Air Monitoring Program (CAMP) stations in selected American cities and in special aerometric studies has revealed some distinct temporal patterns of variations in urban CO levels. There are diurnal, weekly, and seasonal modes. Volume of motor vehicle traffic and meteorological conditions are the dominant factors in such variance. Ambient CO levels in communities correlate well with traffic volume, the highest levels being found most often in places where vehicular traffic is heaviest [23–30]. Similar findings have been reported in studies in large foreign urban communities such as London [28], Paris [29], and Frankfurt [31].

1. *Diurnal Patterns*

While the exact shape of the CO curve is dependent on local traffic patterns, two daily peaks corresponding to the morning and evening traffic 'rush' hours occur in most communities (see Figure 1). The initial daily maxima are found between 7:00 and 9:00 a.m., coincident with heavy morning automobile traffic volume; the second peak is reached in the late afternoon and early evening [26, 27]. Within a community, there is little change in the time of occurrence of the daily morning maximum CO levels during the year [23, 26]. Although a late afternoon rise is evident in all seasons, a very pronounced evening 'rush hour' peak is found only in the winter CO curve. An exception to these general observations may be found in 'downtown' New York City, where there is a rapid rise in the morning CO levels corresponding to the morning motor vehicle 'rush hour' traffic, with a uniformly high plateau lasting until afternoon; then a slower step rise in CO concentrations begins and builds to a single peak in the

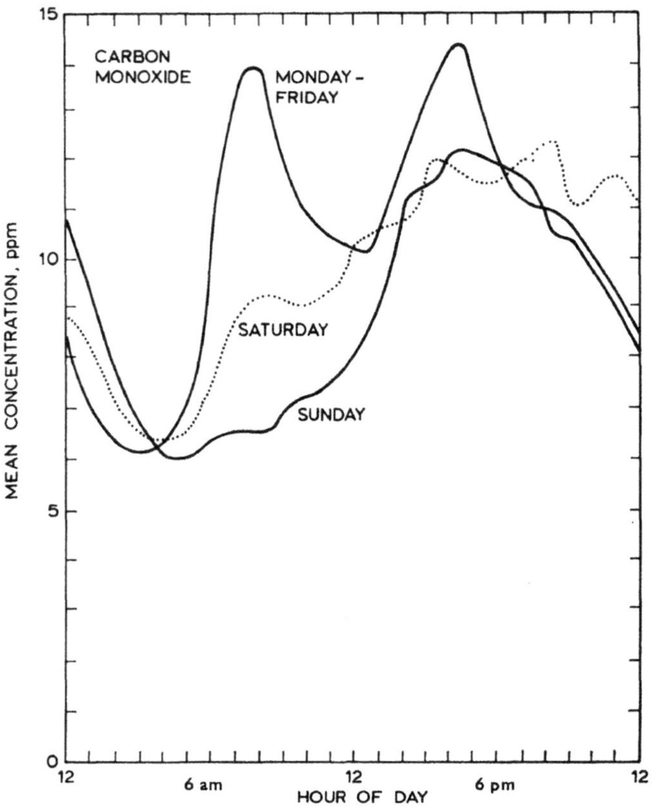

Fig. 1. Diurnal variation of carbon monoxide levels on weekdays, Saturdays, and Sundays in Chicago, 1962–64. (From Ref. [25], Depth. of Commerce.)

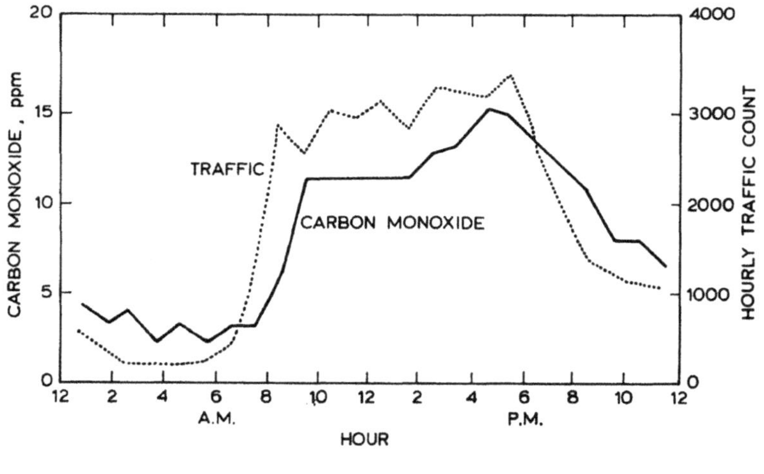

Fig. 2. Hourly average carbon monoxide concentration and traffic count in mid-town Manhattan. (From Johnson *et al.* [30].)

late afternoon [30] (see Figure 2). The shape of this curve is indicative of saturation levels of traffic [30].

2. *Weekly and Seasonal Patterns*

Peak concentrations are higher on weekdays than on Saturdays, which in turn, are higher than on Sundays and holidays, corresponding to the relative traffic volumes [24, 26] (see Figure 1). Distinct seasonal patterns are related to traffic volume, mode of driving, and meteorological variables, the mean community CO concentration being higher generally in autumn than in winter, spring, and summer [26].

B. METEOROLOGICAL FACTORS

The rates of emission and dispersion of CO determine the level at a given location. Both macrometeorological elements, such as atmospheric stability and wind speed, and micrometeorological elements, such as mechanical turbulence, play a role in the rate of dispersion of ambient CO [1, 27].

During prolonged periods of air stagnation, characterized by poor or inadequate diffusion, which occur periodically in most urban communities, the atmospheric levels of CO and other air pollutants will build up. For example, in the fall of 1964, during unusually prolonged and severe inversions, CO concentrations measured at air monitoring stations in Los Angeles and Sacramento, Calif. exceeded 30 ppm (35 mg/m^3) for 8 hour periods [32]. Lawther *et al.* [28] reported a peak community CO level reading in downtown London at street level on a calm day of 235 ppm (270 mg/m^3) in 1957.

C. COMMUNITY CARBON MONOXIDE CONCENTRATIONS

Data on community CO levels have been obtained from continuous air monitoring program (CAMP) stations located in 'downtown', off-street locations in a number of large U.S. cities; commuter traffic surveys; state, regional and municipal aerometric surveys; and special studies. These data indicate that a wide range of ambient CO levels exist within any community with significant localized differences dependent on proximity to traffic, traffic volume, type of traffic, and meteorological variables. These community CO levels range from 1 to >140 ppm, the latter being present only as brief peaks in heavy traffic.

Brice and Roesler [33], in a comparative study of CO levels in several U.S. cities participating in the aforementioned CAMP network, found that mean commuter traffic CO concentrations, based on 30-min integrated samples, were from 1.3 to 6.8 times greater than simultaneous mean aerometric levels measured at the corresponding CAMP station unit. Lynn *et al.* [34], in a study of commuter CO exposures in a number of U.S. cities, found that people in moving vehicles, particularly those in heavy traffic, are at times exposed to sustained levels of 50 or more ppm CO. Such mean CO levels, based on integrated 30-min aerometric samples taken in traffic, were found in the central business areas of 11 out of the 15 large U.S. cities surveyed. Very brief peak CO exposures, however, in all of these cities usually far exceeded 50 ppm and reached as high as 147 ppm in Los Angeles arterial traffic and 141 ppm in New York expressway traffic.

Larsen and Burke [35] using a mathematical model, recently have statistically analyzed extensive aerometric CO data from a variety of sampling sites in many U.S. cities. This analysis, based on maximum annual 8-hour averaging time concentrations at the most polluted 5% of the sites, revealed that ambient CO levels found in heavy traffic in general were more than *twice* the levels found at comparative 'CAMP' stations and about *five* times the CO levels found in residential areas. The CO levels at 'CAMP' stations in 'downtown' areas of major cities, on the average, were about *twice* the concentrations found in residential areas.

Carbon monoxide is also a serious atmospheric pollutant in urban areas of other industrialized countries [28, 29, 31, 38]. Lawther *et al.* [28], for example, reported average CO levels in central London ranging from 10 to 55 ppm over 10 min sampling periods with a mean concentration of 36 ppm for the 8-hour period from 11:00 a.m. to 7:00 p.m. and a maximum 10-minute CO concentration of 155 ppm during the evening rush hour.

D. SPECIAL AREA CARBON MONOXIDE CONCENTRATIONS

While significant concentrations of CO occur in community air of urban areas, particularly in traffic, even higher concentrations often exceeding 100 ppm for sustained periods, have been reported in underground garages and tunnels, and at loading platforms [36–38]. Waller *et al.* [37], for example, found the mean hourly CO levels in the Blackwall Tunnel in London in 1958–59 generally averaged >100 ppm during rush hours with the mean hourly concentration reaching as high as 295 ppm on some days during the morning rush hours. Mean hourly concentrations, on the other hand, in the Sumner Tunnel in Boston usually were less than 100 ppm except for the evening rush hours when an hourly mean of 126 ppm was reached [36].

E. CHEMICAL REACTIONS OF CARBON MONOXIDE

While CO is both formed and oxidized in the upper atmosphere, the rate of chemical oxidation of CO in the dense lower atmosphere is very slow and unimportant [10, 11]. It is essentially chemically inert, apparently not reacting with other con-constituents of urban air to a significant degree.

1. *Lower Atmospheric Reactions*

Carbon monoxide is chemically oxidized very slowly in the dense lower atmosphere. Two such oxidation reactions, however, do occur [39, 40]:

$$CO + O_2 \rightarrow CO_2 + O, \tag{1}$$

and in the presence of moisture

$$CO + H_2O \rightarrow CO_2 + H_2. \tag{2}$$

Both of these reactions have appreciable energy barriers, 51 and 56 kcal per mole (kcal/mole), respectively. Direct chemical reactions between CO and oxygen or water occur at a frequency of less than one reaction per 10^{15} molecular collisions at room temperature. These reactions become important as gas-phase processes primarily at temperatures above 500 °C and even more so above 1000 °C. These reactions occur more readily and at lower temperatures on the surfaces of certain catalysts, usually metal oxides. Indeed, this oxidation reaction is catalyzed at room temperatures by metallic catalysts, such as the oxides of magnesium and copper, and by palladium on silica gel [41].

Ozone will oxidize CO to CO_2, but the rate of this reaction is extremely slow at atmospheric temperatures and concentrations [42, 43]. A high activation energy of about 20 kcal for the oxidation of CO by ozone has been found [44].

Oxidation by nitrogen dioxide in the reaction:

$$NO_2 + CO \rightarrow CO_2 + NO \tag{3}$$

has an even higher activation energy than does the oxidation of CO by ozone [45]. The activation energy of 28 kcal essentially rules out the occurrence of this process in the lower atmosphere.

Consideration has also been given to the possibility that some very rapid reactions may occur between CO and certain intermediates of photochemical smog reactions. One possible intermediate is the hydroxyl (OH) radical, which can be produced by the photolysis of aldehydes to produce perhydroxyl, and then be reduced to the hydroxyl radical. The presence of free OH radicals in the dense lower atmosphere, however, is still generally conjectured primarily because the chemistry of OH pertinent to such systems is largely unexplored [11].

Carbon monoxide, in addition to combining with the OH radical also reacts rapidly with other free radicals. In spite of the high intrinsic speed of these reactions in the laboratory, their importance is diminished by the low concentrations of these free radicals in the normal atmosphere. Although future research with hydroxyl class radicals may be rewarding, at present one must conclude that there are no proven significant gaseous oxidation reactions of CO in the ambient lower atmosphere.

2. *Upper Atmospheric Reactions*

Short wavelength ultraviolet radiation (about 1700 Å) dissociates CO_2 into CO and atomic oxygen according to the reaction [10]:

$$CO_2 + h\nu \rightarrow CO + O\,. \tag{4}$$

The yield of CO from the photodissociation of CO_2 in the upper atmosphere, however, is considered to be relatively small at levels below 100 km, since the intensity of active ultraviolet radiation falls off rapidly at that level.

Carbon dioxide may be reformed by a three-body collision:

$$CO + O + M \rightarrow CO_2 + M\,, \tag{5}$$

where M represents the third body. The probability of this occurring is small, however, because of the infrequency of such third bodies at these altitudes [46].

4. Global Emissions, Background Levels, and Fate of Atmosphere Carbon Monoxide

A. GLOBAL EMANATIONS

Global emissions of CO from technological sources and forest fires have heretofore been estimated to approximate 230 million tons per year [3]. There has been a steady increase in combustion and other sources which release CO to the atmosphere [47]. Emissions of CO in the U.S.A. alone are expected to double in the period from 1960 to 1980 unless major reductions are made in automobile exhaust emissions. On a global basis, the increase would be expected to be just as significant because of the rapid technological changes and the large increase in the number of motor vehicles in other countries. Due to this increase in CO from technological sources and with the addition from oceanic sources [22], it is now estimated that about 250 million tons of CO are emitted globally every year.

B. BACKGROUND LEVELS

The amount of CO measurable in relatively unpolluted ('clean') air is small. Junge reported that the background level of CO in the lower atmosphere is in the range of 0.01 to 0.2 ppm [48], based on a limited number of infrared solar spectra taken at Mt. Wilson, California; Ottawa, Canada; Jungfraujoch, Switzerland; and Columbus, Ohio, with an average concentration of about 0.1 ppm; there appears to be no significant variation in CO levels with geographical location [49]. Robbins *et al.* [50] determined that North Pacific marine air contained as little as 0.025 ppm while the non-urban air mass over continental California contained levels of 0.05 to 1.0 ppm CO. Recent studies in northern Alaska indicate a range of 0.055 to 0.260 ppm, averaging 90 ppb (0.09 ppm) [51]. Robbins *et al.* also measured typical background levels of CO ranging from 0.24 to 0.90 ppm at Camp Century, Greenland, and concluded that the variability of CO in unpolluted areas is a characteristic of the air mass in transit and reflects the prior history of the air mass.

C. ESTIMATED MEAN LIFETIME OF ATMOSPHERIC CARBON MONOXIDE

Carbon monoxide is a relatively long-lived substance in the lower atmosphere. The precise mean residence time of CO (r), however, is not known with certainty. Estimates of the CO mean residence time in the troposphere range from a lower limit of 0.1 year [52, 53] to about 5 years [50]. Based on the ongoing studies of CO in ice samples and in aerometric monitoring at Camp Century, Robbins *et al.* [50] have tentatively concluded that the background levels of CO do not appear to be rising at the present time.

D. FATE OF CARBON MONOXIDE IN THE ATMOSPHERE

In the absence of scavenging processes, the revised estimated world-wide emissions of CO from both technological and natural sources, on an order of about 250×10^6 tons/year, would be sufficient to raise the atmospheric background 0.04 ppm/year (extrapolated from Ref. [54]).

Based on the large mass of CO generated and on the continued increases in CO emissions within recent years and with the assumption of the absence of any sinks, it is estimated that the current average global background levels would have well exceeded 1 ppm. Nevertheless, the background levels have not risen and have remained relatively constant. The relative inertness of CO with the normal gaseous atmospheric constituents and its photon transparency [46], effectively eliminate the possibility of chemical oxidation as a mechanism for CO removal in the lower atmosphere (except for the possible reaction with OH). Because of the aforementioned anomaly, it has, therefore, been postulated that some other sink or scavenging process for removal of atmospheric CO exists. A number of potential removal processes have recently been reviewed by Jaffe [1, 6].

5. Possible Removal Processes for Carbon Monoxide

A. ATMOSPHERIC MIGRATION (UPPER ATMOSPHERIC SINK)

Carbon monoxide in the lower atmosphere conceivably may eventually migrate by atmospheric diffusion, mixing and turbulence to the upper atmosphere where it is oxidized to CO_2 in the presence of high intensity, ultraviolet solar radiation. A recent laboratory study by Harteck and Reeves [55] has confirmed that CO, in the presence of NO_2 or other absorbing molecules such as O_3, when subjected to high-intensity, ultraviolet radiation in an evacuated chamber, is oxidized to CO_2.

B. BIOLOGICAL REMOVAL (TERRESTRIAL AND MARINE BIOSPHERE SINK)

Another possible removal mechanism of atmospheric CO is the presence, in significant numbers, of microorganisms and plants that can metabolize CO.

(a) *Soil bacteria.* The earth's surface is a possible agent in the removal of CO from the atmosphere. Carbon monoxide in contact with the soil may be oxidized to CO_2 or converted to methane (CH_4) by common specific anaerobic methane-producing soil microorganisms, *Methanosarcina Barkerii* and *Methanobacterium formicum*, in the presence of moisture.

This action has been demonstrated in the laboratory by Schnellen [56], who showed that pure cultures of these bacteria utilize CO as a source of carbon and convert CO into methane. Schnellen found that *Ms. Barkerii* is capable of effecting a considerable conversion of CO to CH_4 according to the equation:

$$4CO + 2H_2O \rightarrow CH_4 + 3CO_2 . \tag{6}$$

Stephenson [57], however, indicates that CO, in the absence of H_2, reacts with water in these bacteria in two stages as follows:

$$4CO + 4H_2O \rightarrow 4CO_2 + 4H_2 , \tag{7}$$

and

$$CO_2 + 4H_2 \rightarrow CH_4 + 2H_2O . \tag{8}$$

In the presence of H_2, these bacteria convert CO directly into methane and water:

$$CO + 3H_2 \rightarrow CH_4 + H_2O . \tag{9}$$

An aerobic soil bacterium, *Bacillus oligocarbophilus* (*Carboxydomonas oligocarbophila*), found in and isolated from arable soil, has also been demonstrated to utilize CO as a source of carbon by oxidizing CO to CO_2. The organism, when cultivated on simple organic media free from other carbon sources, oxidizes CO to CO_2, which is then utilized as a source of energy [58, 59]. Another bacterium, *Clostridium welchii*, when grown in the presence of CO, has been reported to produce lactic acid as a fermentation product [60].

Although these soil bacteria utilize CO in their fermentation or metabolic processes, it is difficult to estimate global destruction rates of CO on the basis of such laboratory experiments inasmuch as the prevalence and world-wide distribution of these bacteria in soils and in the biosphere are presently unknown.

(b) *Absorption or Retention by Vegetation.* The process of plant respiration may also serve as a potential CO removal process, but this is not firmly established. Plants are known to be scavengers for a wide variety of atmospheric materials. Although CO is not toxic to vegetation at concentrations found in ambient air, nevertheless, a number of reactions have been noted in plants after exposure to relatively high concentrations of CO: Ducet and Rosenberg [61] report the inhibition of respiration processes by CO, Burris [62] cites the inhibition of nitrogen fixation by CO, and Carr [63] has observed a variety of visible changes when plants are exposed to CO. Evidence of the fact that CO is a phytotoxicant at high concentrations reinforces the speculation that CO entering a plant during the respiration process will undergo subsequent reaction within the plant.

C. BIOCHEMICAL REMOVAL (BIOCHEMICAL SINK)

A potential biochemical removal process for CO is the binding of CO to the porphyrin-type compounds that are widely distributed in plants and animals. In particular, the heme compounds, such as hemoglobin found in man and animals, which are analogous to porphyrin compounds found in plants, are known to bind CO. It must be noted, however, that practically all of the CO absorbed by these heme compounds is eventually discharged from the blood of man and animals and only a small fraction is retained [64]. Nevertheless, this type of process in vegetation may have important potential for scavenging atmospheric CO. Permanent removal from the environment, however, would depend on whether CO subsequently entered into some reaction process to form CO_2 when the porphyrin compound is degraded.

D. OCEANIC SINK

While absorption in the world's oceans is a recognized sink for atmospheric CO_2, there is no evidence at present that the oceans are a sink for CO since no process or reaction has been discovered that would remove CO from the atmosphere [1, 3]. Pure gaseous carbon monoxide at 1 atm pressure is soluble to a limited degree in sea water, on the order of 32 to 17 ml gas STPD/liter of water over a surface water temperature range of $-2°$ to $+30°C$ and a chlorinity range of 15 to 21, respectively [65] (convertible to salinities of 27.1 to 37.9, respectively). As an illustration, at 25 °C, the solubility of CO is 18 ml/liter of sea water (volume of gas (STPD)) absorbed by a unit volume of water when the pressure of the gas equals 1 atm (760 mm) at a salinity of 36 parts per thousand [59]. This solubility rate is roughly between the solubilities of O_2 and N_2, but is considerably less than the solubilities of CO_2 and SO_2 in water. (The solubility coefficients of O_2, N_2 and CO in sea water and in pure (distilled) water differ by about 10%.) The precise degree of solubility, however, is a function of the partial pressure of the gas in the atmosphere. Thus, based on an atmospheric CO background level of 0.01 to 0.2 ppm as found in clean air regions, the solubility of CO in sea water will be 1.8×10^{-7} to 3.6×10^{-6} ml CO/liter of water at equilibrium between the atmosphere and surface waters, assuming that the atmosphere is the sole source of the gas.

Swinnerton *et al.* [66] recently simultaneously measured the CO content of the atmosphere and of the surface water at 29 different points during an oceanographic cruise between Washington, D.C. and Puerto Rico. These investigators found that the actual measured CO concentration of the surface waters (W) at all sampling points ranged from 7 to about 90 times (averaging about 28 times) the aforementioned theoretical CO concentrations of the surface water (T) based on the concentration of the gas in the water if the only source were the atmosphere. These data indicate that atmospheric CO from land sources may not be the principal source of CO in the water since the highest values of W/T were found in the open water. Marine biological sources of CO such as marine algae and siphonophores and/or other marine sources apparently contribute substantial quantities of CO to the surface waters exceeding that obtained from the atmosphere. These findings [66], confirmed in a more recent study [22], indicate that the ocean in the areas studied is not a sink for atmospheric CO but, indeed, serves as an additional natural source, with the transient exchange or net transport of CO being from the water to the atmosphere due to the supersaturated condition of the surface waters with respect to the partial pressure of CO in the atmosphere over the ocean [22]. Additional studies are being conducted. This evidence, however, does not preclude the possibility of the ocean serving as a sink for atmospheric CO as well. In the case of CO_2, it has been demonstrated that the oceans are a major sink for atmospheric CO_2 as well as a source for release to the atmosphere [67].

E. ADSORPTION ON SURFACES

The catalytic adsorption of CO on hot metallic surfaces is well known. Adsorption of CO on copper surfaces, on charcoal, on Pyrex glass, and on quartz have been reported in laboratory studies at temperatures of 300 °C and above [68–72]. Kummler *et al.* [52] indicate that the gas-phase reaction of CO with nitrous oxide (N_2O), which is considered to be too slow to be of importance in the atmosphere, is, nevertheless, catalyzed in the presence of certain surfaces such as charcoal, carbon black, and glass in laboratory exposures of 300 °C and above, wherein the CO is oxidized to CO_2.

$$CO + N_2O \xrightarrow[\text{surface}]{} CO_2 + N_2 . \qquad (10)$$

These investigators have extrapolated the reported reaction rates to lower temperatures (300 K, ≅ 27 °C) such as found in the lower atmosphere, and consider that such a reaction of the two gases in the presence of such surfaces is feasible in the lower atmosphere [52]. This conclusion is supported in part by laboratory studies wherein Gardner and Petrucci [72] measured and observed the chemisorption of CO on metallic films such as copper, cobalt, and nickel oxides at room temperature by means of infrared spectroscopy.

The necessary data for evaluation of the catalytic efficiency of such commonly found surfaces as metals, soils and atmospheric particulates in the adsorption of CO at realistic ambient temperatures found in the lower atmosphere, however, are presently unavailable. Hence, the possibility of such commonly found surfaces serving as a significant sink for atmospheric CO by adsorption is unclear and presently uncertain.

6. Summary

Carbon monoxide, principally a man-made pollutant, is the most abundant air pollutant found in the lower atmosphere. Total emissions from technological sources alone generally exceed those of all other air pollutants combined (excluding CO_2), with the total tonnage in the U.S.A. alone estimated to be 102 million tons. Fuel combustion by motor vehicles is by far the principal source. Other important technological sources of CO are industrial processes, aircraft exhaust, solid waste combustion, miscellaneous man-made fires and stationary fuel combustion. Several natural sources have also been described including wild ('unprescribed') forest fires.

The CO concentrations in metropolitan areas are quite variable and localized, being dependent on human activity and meteorological factors. There are diurnal, weekly, and seasonal patterns of CO variation. Community CO concentrations range from 1 to about 140 ppm, the latter high levels being present only as very brief peaks in heavy traffic. Mean carbon monoxide levels (based on integrated 30-min samples) of 50 ppm have been measured in heavy traffic in many U.S. cities. Background levels of CO in remote areas, on the other hand, range from 0.01 to about 1 ppm; the variability in non-polluted areas appears to be a characteristic of the air mass intransit.

Global emission levels of CO from technological and natural sources are estimated to be about 250 million tons per year. With the mean residence time of atmospheric CO estimated to range from 0.1 year to about 5 years, it is further estimated that the background levels would increase 0.04 ppm per year. On-going studies in Greenland, however, indicate that the background levels of CO are not increasing. Inasmuch as the chemical oxidation of CO to CO_2 in the lower atmosphere is very slow, it has been postulated that one or more removal processes or sinks for CO exist.

A number of potential sinks have been discussed, including upper atmospheric migration, biological metabolism by terrestrial and/or marine organisms, biochemical combination with porphyrin compounds, oceanic absorption, and adsorption on various surfaces. The precise removal mechanism, however, is presently unknown.

References

[1] Jaffe, L. S.: 1968, 'Ambient Carbon Monoxide and its Fate in the Atmosphere', *J. Air Pollution Control Assoc.* **18** (8), 534–540.

[2] Air Quality Criteria for Carbon Monoxide: 1970, National Air Pollution Control Administration Publication No. AP–62. Environmental Health Service, Public Health Service. U.S. Department of Health, Education and Welfare, Washington, D.C., March 1970.

[3] Robinson, E. and Robbins, R. E.: 1968, Sources, Abundance and Fate of Gaseous Atmosphere Pollutants. Stanford Research Institute Project Pr–6755. Menlo Park, Calif.

[4] Grob, K.: 1968, 'Gaseous Components of Tobacco Smoke', in *Toward a Less Harmful Cigarette*, Monograph No. 28. National Cancer Institute. National Cancer Institute, Bethesda, Md., p. 215.

[5] Federal Trade Commission: 1969, Personal communication.

[6] Jaffe, L. S.: 1970, 'Sources, Characteristics and Fate of Atmospheric Carbon Monoxide', Paper presented at the Conference on Biological Effects of Carbon Monoxide, New York Academy of Sciences, January 1970.

[7] Flury, F. and Zernik, F.: 1931, *Schädliche Gase, Dämpfe, Nebel, Rauch and Staubarten*, Julius Springer, Berlin.

[8] White, J. J.: 1932, 'Carbon Monoxide and its Relation to Aircraft', *U.S. Naval Med. Bull.* **30**, 151.

[9] Migeotte, M.V. and Neven, L.: 1952, 'Recent Progress in the Observation of the Solar Spectrum of Jungfraujoch', *Mem. Soc. Roy. Sci. Liège* **12**, 165.

[10] Bates, D. R. and Witherspoon, A. E.: 1952, 'The Photochemistry of some Minor Constituents of the Earth's Atmosphere', *Monthly Notices Roy. Astron. Soc.* **112**, 101–124.

[11] Leighton, P. A.: 1961, 'Photochemistry of Air Pollution, IX', in *Phys. Chem.*, A Series of Monographs, Academic Press, New York.

[12] Altshuller, A. P. and Bufalini, J. J.: 1965, 'Photochemical Aspects of Air Pollution: a Review', *Photochem. Photobiol.* **4**, 97–146.

[13] Wilks, S. S.: 1959, 'Carbon Monoxide in Green Plants', *Science* **129**,964–966.

[14] Siegel, S. M., Renwick, G., and Rosen, L. A.: 1962, 'Formation of Carbon Monoxide during Seed Germination and Seedling Growth', *Science* **137**,683–684.

[15] Loewus, M. W. and Delwiche, C. C.: 1963, 'Carbon Monoxide Production by Algae', *Plant Physiol.* **38**(4), 371–374.

[16] Chapman, D. J. and Tocher, R. D.: 1966, 'Occurrence and Production of Carbon Monoxide in some Known Algae', *Can. J. Botany* **44**,1438–1442.

[17] Westlake, D. W., Roxburgh, J. M., and Talbot, G.: 1961, 'Microbial Production of Carbon Monoxide from Flavonoids', *Nature* **189**, 510–511.

[18] Barham, E. G.: 1963, 'Siphonophores and the Deep Scattering Layer', *Science* **140**, 826–828.

[19] Barham, E. G. and Wilton, J. W.: 1964, 'Carbon Monoxide Production by a Bathypelagic Siphonophore', *Science* **144**, 860–862.

[20] Wittenberg, J. B.: 1960, 'The Source of Carbon Monoxide in the Float of *Physalia Physalis*, the Portuguese Man-of-War', *J. Exp. Biol.* **37**, 698–705.

[21] Coburn, R. F., Blakemore, W. S., and Forster, R. E.: 1963, 'Endogenous Carbon Monoxide Production in Man', *J. Clin. Invest.* **42**, 1172–1178.

[22] Swinnerton, J. W., Linnenbom, V. J., and Lamontague, R. A.: 1970, 'The Ocean: A Natural Source of Carbon Monoxide', *Science* **167**, 984–987.

[23] Dickinson, J. E.: 1961, *Air Quality of Los Angeles County*. Technical Progress Report, vol. II. Los Angeles Air Pollution Control District, Los Angeles, Calif.

[24] Brief, R. S., Jones, A. R., and Yoder, J. S.: 1960, 'Lead, Carbon Monoxide, and Traffic, a Correlation Study', *J. Air Pollution Control Assoc.* **10**, 384–388.

[25] Department of Commerce: 1967, The Automobile and Air Pollution: A Program for Progress, Part II. Subpanel reports to the panel on electrically powered vehicles. Commerce Technical Advisory Board. U.S. Government Printing Office, Washington, D.C.

[26] Department of Health, Education and Welfare: 1966, Continuous Air Monitoring Projects. 1962–1967 Summary of monthly means and maximums. National Air Pollution Control Administration Publication No. APTD 69-1. Public Health Service, Arlington, Va.

[27] McCormick, R. A. and Xintaras, C.: 1962, 'Variation of Carbon Monoxide Concentrations as Related to Sampling Interval, Traffic and Meteorological Factors', *J. Appl. Meteorol.* **1** (2), 237–243.

[28] Lawther, P. J., Commins, B. T., and Henderson, M.: 1962, 'Carbon Monoxide in Town Air: an Interim Report', *Ann. Occupational Hyg.* **5**, 241–248.
[29] Chovin, P.: 1967, 'Carbon Monoxide: Analyses of Exhaust Gas Investigations in Paris', *Envir. Res.* **1**, 198–216.
[30] Johnson, K. L., Dworetsky, L. H., and Heller, A. N.: 1968, 'Carbon Monoxide and Air Pollution from Automobile Emissions in New York City', *Science* **160** (3823), 67–68.
[31] Georgii, H. W. and Weber, E.: 1962, 'Untersuchung der Kohlenoxyd-emission in einer Grosstadt' [Investigations of Carbon Monoxide Emissions in a Large City], *Intern. J. Air Water Pollution* **6**, 179–195.
[32] Tebbens, B. D.: 1968, 'Gaseous Pollutants in the Air', in *Air Pollution*, vol. I (ed. by A. C. Stern), Second Edition, Academic Press, New York, Ch. 2, pp. 31–32.
[33] Brice, R. M. and Roesler, J. F.: 1966, 'The Exposure to Carbon Monoxide of Occupants of Vehicles Moving in Heavy Traffic', *J. Air Pollution Control Assoc.* **16**, 597–600.
[34] Lynn, D. A., Tabor, E., Ott, W., and Smith, R.: 1967, Present and Future Commuter Exposures to Carbon Monoxide. Paper No. 67-5, presented at the 60th Annual Meeting, Air Pollution Control Association, Cleveland, Ohio.
[35] Larsen, R. I. and Burke, H.: 1969, Ambient Carbon Monoxide Exposures. Paper 69-167, presented at the 62nd Annual Meeting Air Pollution Control Association, New York, June 1969.
[36] Conlee, C. J., Kenline, P. A., Cummins, R. L., and Konopinski, V. J.: 1967, 'Motor Vehicle Exhaust at Three Selected Sites', *Arch. Environ. Health* **14**, 429–446.
[37] Waller, R. E., Commins, B.T., and Lawther, P. J.: 1961, 'Air Pollution in Road Tunnels', *Brit. J. Ind. Med.* **18**, 250–259.
[38] Trompeo, G., Turletti, G., and Giarrusso, O. T.: 1964, 'Concentrations of CO in Underground Garages', *Rass. Med. Ind.* **33**, 392–393.
[39] Fischer, E. R. and McCarthy, M. Jr.: 1966, 'Study of Reaction of Electronically Excited Oxygen Molecules with Carbon Monoxide', *J. Chem. Phys.* **45**, 781–784.
[40] Grave, W. M. and Long, F. J.: 1954, 'Kinetics and Mechanisms of the two Opposing Reactions of the Equilibrium: $CO + H_2O = CO_2 + H_2$', *J. Am. Chem. Soc.* **76**, 2602–2607.
[41] The Merck Index of Chemicals and Drugs, Seventh Edition, 1960: *Carbon Monoxide*, (ed. by P. G. Stecher), Merck and Company, Inc., Rahway, N. J., p. 212.
[42] Garvin, D.: 1954, 'The Oxidation of Carbon Monoxide in the Presence of Ozone', *J. Am. Chem. Soc.* **76**, 1523–1527
[43] Harteck, P. and Dondes, S.: 1957, 'Reactions of Carbon Monoxide and Ozone', *J. Phys. Chem.* **26**, 1734–1737.
[44] Zatsiorskii, M., Kondrateev, V., and Solnishkova, S.: 1940, 'Izluchenie plameni $CO + O_3$ i mekhanism etoi reaktsii' [Radiation of the flame of $CO + O_3$ and the mechanism of this reaction], *Zn. Fiz. Kim.* **14**, 1521–1527.
[45] Brown, F. B. and Crist, R. H.: 1941, 'Further Studies on the Oxidation of Nitric Oxides: The Rate of Reaction between Carbon Monoxide and Nitrogen Dioxide', *J. Chem. Phys.* **9**, 840–846.
[46] Penndorf, R.: 1949, 'The Vertical Distribution of Atomic Oxygen in the Upper Atmosphere', *J. Geophys. Res.* **54**, 1–38.
[47] Heller, A. N. and Walters, D. F.: 1965, 'Impact of Changing Patterns of Energy Use on Community Air Quality', *J. Air Pollution Control Assoc.* **15**, 423–428.
[48] Junge, C. E.: 1963, *Air Chemistry and Radioactivity*, Academic Press, New York.
[49] Locke, J. L. and Herzberg, L.: 1953, 'The Absorption due to Carbon Monoxide in the Infrared Solar Spectrum', *Can. J. Phys.* **31**, 504–516.
[50] Robbins, R. C., Borg, K. M., and Robinson, E.: 1968, 'Carbon Monoxide in the Atmosphere' *J. Air Pollution Control Assoc.* **18**, 106–110.
[51] Cavanagh, L. E., Schadt, C. F., and Robinson, E.: 1969, 'Atmospheric Hydrocarbons and Carbon Monoxide Measured at Point Barrow, Alaska', *Environ. Sci. Technol.* **3**, 251–257.
[52] Kummler, R. H., Grenda, R. N., Baurer, T., Bortner, M. H., Davis, J. H., and MacDowall, J.: 1969, 'Satellite Solution of the Carbon Monoxide Sink Anomaly'. Paper presented at the 50th Annual Meeting of the American Geophysical Union, Washington, D.C. (*J. Geophys. Res.*, in press).
[53] Weinstock, B.: 1969, 'Carbon Monoxide: Residence Time in the Atmosphere', *Science* **166**, 224–225.

[54] Haagen-Smit, A. J. and Wayne, L. G.: 1968, 'Atmospheric Reactions and Scavenging Processes', Chapter 6, in *Air Pollution*, vol. I (ed. by A. C. Stern), 2nd Edition, Academic Press, New York, p. 181.
[55] Harteck, P. and Reeves, R. R., Jr.: 1967, Some Specific Photochemical Reactions in the Atmosphere'. Paper No. 20. Presented at the symposium on the Chemistry of the Natural Atmosphere. American Chemical Society. 154th Annual Meeting. Chicago, Ill.
[56] Schnellen, C. G.: 1947, *Onderzoekingen over de methaangisting*, Doctoral thesis, Technische Wetenschap, Delft, Rotterdam (The Netherlands).
[57] Stephenson, M.: 1949, *Bacterial Metabolism*, Third edition, Longmans, Green and Company, New York City.
[58] Kaserer, H.: 1906, 'Die Oxydation des Wasserstoffes durch Microorganismen', *Zentr. Bakteriol.*, Parasitenk., Abt. **11** (16), 681–696.
[59] Rabinovitch, E. I.: 1945, *Photosynthesis and Related Processes*, Interscience Publishers, New York.
[60] Waksman, S. A.: 1929, *Principles of Soil Microbiology*, Williams and Wilkins Company, Baltimore.
[61] Ducet, G. and Rosenberg, A. J.: 1962, 'Leaf Respiration', *Ann. Rev. Plant Physiol.* **13**, 171–200.
[62] Burris, R. H.: 1966, 'Biological Nitrogen Fixation', *Ann. Rev. Plant Physiol.* **17**, 155–184.
[63] Carr, D. J.: 1961, 'Chemical Influences of the Environment', in *Encycl. Plant Physiol.* **16**, 773–775.
[64] Tobias, C. A., Lawrence, J. H., Roughton, F. J. W., Root, W. S., and Gregerson, M. I.: 1945, 'The Elimination of Carbon Monoxide from the Human Body with Possible Conversion of CO to CO_2, *Am. J. Physiol.* **145**, 253–263.
[65] Douglas, E.: 1967, 'Carbon Monoxide Solubilities in Sea Water', *J. Phys. Chem.* **71**, 1931–1933.
[66] Swinnerton, J. W., Linnenbom, V. J., and Cheek, C. H.: 1969, 'Distribution of Methane and Carbon Monoxide between the Atmosphere and Natural Waters', *Environ. Sci. Technol.* **3** (9), 836–838.
[67] Skirrow, G.: 1965, 'The Dissolved Gases – Carbon Dioxide', in *Chemical Oceanography*, vol. I (ed. by J. P. Riley and G. Skirrow), Academic Press, New York, pp. 312–317, Ch. 7.
[68] Madley, D. G. and Strickland-Constable, R. F.: 1953, 'The Kinetics of the Oxidation of Charcoal with Nitrous Oxide', *Trans. Faraday Soc.* **49**, 1312–1324.
[69] Smith, R. N. and Mooi, J.: 1955, 'The Catalytic Oxidation of Carbon Monoxide by Nitrous Oxyde on Carbon Surfaces', *J. Phys. Chem.* **59**, 814–819.
[70] Strickland-Constable, R. F.: 1938, 'Part played by Surfaces Oxides in the Oxidation of Carbon', *Trans. Faraday Soc.* **34**, 1074–1080.
[71] Krouse, A.: 1961, 'Mechanism of Catalytic Oxidation of CO with N_2O', *Bull. Acad. Polon. Sci., Ser. Sci. Chem.* **9**, 5.
[72] Gardner, R. A. and Petrucci, R. H.: 1960, 'The Chemisorption of Carbon Monoxide on Metals', *J. Am. Chem. Soc.* **82**, 5051–5053.

For Further Reading

1. A. C. Stern (ed.), *Air Pollution*, Second Edition, Vols. I and II, Academic Press, New York, 1968.
2. Cleaning our Environment: The Chemical Basis for Action, 1969. A Report by the Subcommittee on Environmental Improvement. Committee on Chemistry and Public Affairs, American Chemical Society, Washington, D.C.
3. Restoring the Quality of Our Environment, November 1965. Report of the Environmental Pollution Panel, President's Advisory Committee, The White House, Washington, D.C.
4. W. Ott, J. F. Clarke, and G. Ozolins, Calculating Future Carbon Monoxide Emissions and Concentrations from Urban Traffic Data. Public Health Service Publication No. 999-AP-41, Department of Health, Education and Welfare, Washington, D.C., 1967.

GASEOUS ATMOSPHERIC POLLUTANTS FROM URBAN AND NATURAL SOURCES*

ELMER ROBINSON and ROBERT C. ROBBINS
Stanford Research Institute, Menlo Park, Calif., U.S.A.

Abstract. Major aspects of the circulation through the atmospheric environment of a number of gaseous pollutants have been estimated, including source magnitudes, residual atmospheric concentrations, and scavenging processes. The compounds considered include the major sulfur and nitrogen pollutants as well as CO. One-third of the sulfur reaching the atmosphere comes from pollutant sources, mainly as SO_2. Within the atmosphere there is a net transfer of sulfur from land to ocean areas. In the global atmospheric nitrogen cycle, pollutant emissions of NO_2 play only a minor role, and the atmospheric nitrogen cycle is apparently dominated by natural emissions of ammonia. Scavenging mechanisms for removing CO from the atmosphere are very speculative, but there is now considerable evidence of an oceanic source of natural CO to add to the very large pollutant source.

1. Introduction

The atmosphere is a complex chemical system in which the emissions from urban pollution sources mix with emanations from the natural environment. By considering both the pollutant and the natural sources it is possible to improve our understanding of the impact of air pollutants on the atmospheric environment.

This discussion will cover a number of the common atmospheric pollutants: SO_2, NO_2, and CO and their counterparts from the natural environment. We will look at these compounds on an integrated global basis, and will consider the sources of the pollutants, their atmospheric concentrations, their reactions, and available scavenging mechanisms.

This discussion is a brief summary of a portion of the research carried out for the American Petroleum Institute at Stanford Research Institute. Substantiating calculations and additional discussion for much of the material presented in this paper are given in the basic research report published by the API (see Robinson and Robbins, 1968a).

2. SO_2 and the Sulfur Cycle

The sulfur compounds in the atmosphere come from both the natural environment and from air pollution emissions. Natural sulfur compound emissions are SO_4 aerosols** produced in sea spray, and H_2S from the decomposition of organic matter in swamp areas, bogs, and tidal flats. Areas of volcanic activity are also a minor source of H_2S. The emissions of SO_2 come almost exclusively from pollution sources. Some H_2S is also of industrial origin.

* Portions of this paper will appear in the *Journal of the Air Pollution Control Association.*

** For convenience the notation SO_4 will be used in this presentation with the understanding that this is the sulfate ion and that it is actually present as a compound such as $(NH_4)_2SO_4$ or H_2SO_4.

A. SO_2 SOURCES

Annual worldwide pollution emissions of SO_2 have been estimated to be 147×10^6 tons. Of this total, 70% is estimated to result from coal combustion and 16% from the combustion of petroleum products, mainly residual fuel oil. As Table I shows, the remaining tonnage is accounted for by petroleum refining and nonferrous smelting. These estimates are based on 1965 world data.

TABLE I

Estimated global emissions of sulfur compounds

Compound	Source	Estimated emissions (tons/yr)	Emissions as sulfur (tons/yr)
SO_2	Coal combustion	102×10^6	51×10^6
	Petroleum refining	6×10^6	3×10^6
	Petroleum combustion	23×10^6	11×10^6
	Smelting operations	16×10^6	8×10^6
H_2S	Industrial emissions	3×10^6	3×10^6
	Marine emissions	30×10^6	30×10^6
	Terrestrial emissions	70×10^6	70×10^6
SO_4	Marine emissions	130×10^6	44×10^6
		Total emissions	220×10^6

Previous estimates of world SO_2 emissions have been made by Katz (1958). For the years of 1937 and 1940, total emissions were about 69×10^6 tons and 78×10^6 tons, respectively. Thus, SO_2 emissions have roughly doubled in the period between 1940 and 1965.

The estimates for SO_2 emissions have been combined with estimates of H_2S and SO_4 in Table I to show the estimated world natural and pollution emissions of sulfur compounds. The total emission, expressed as sulfur, is 220×10^6 tons. This estimate is not precise, because about 50% of the total comes from estimates of H_2S emissions from land and ocean areas.

B. ATMOSPHERIC REACTIONS AND SCAVENGING PROCESSES

There has been considerable interest in SO_2 scavenging reactions for many years, and much research has centered around the need for specific catalysts to promote SO_2 oxidation in liquid droplets. However, a more realistic process for foggy atmospheres involves ammonia. Junge and Ryan (1958) found that SO_2 had low solubility in water droplets of low pH, but that ammonia promoted the solubility of SO_2 by neutralizing the acid in the droplets formed by the absorbed SO_2. Extrapolation of laboratory experiments to realistic atmospheric conditions indicates that SO_2 lifetimes in foggy conditions might be as short as one hour. The fact that ammonium sulfate is commonly identified in atmospheric particulate samples lends support to this scavenging reaction.

Rainout processes within clouds and washout resulting from falling rain may also be quite effective in scavenging SO_2 as Bielke and Georgii (1967) have shown.

Sulfur dioxide oxidation is not confined to fog or rain conditions. While direct oxidation by molecular oxygen has been shown to be insignificant, photochemical oxidation of SO_2 in mixtures with NO_2 and hydrocarbons is probably one of the more significant scavenging systems for SO_2. The resultant aerosol formed by this system is H_2SO_4. The reaction can proceed with very low concentrations of the constituents (Renzetti and Doyle, 1960).

An integrated system to explain the reaction of SO_2 in the atmosphere is not now available. Under daytime, low humidity conditions the photochemical processes that form H_2SO_4 or sulfate aerosols seem to be most important. At night and with high humidity, fog, or rain, absorption into water drops with subsequent oxidation to SO_4 is probably the most important process.

Sulfur dioxide is also scavenged from the atmosphere by vegetation. In vegetation scavenging it is possible to consider the rate of deposition by using Chamberlain's (1960) concept of a deposition velocity. On the basis of chamber studies of SO_2 intake by vegetation (Katz and Ledingham, 1939), the calculated deposition velocity for SO_2 is about 1 cm/sec. For a concentration of 1 ppb this deposition velocity predicts an SO_2 deposition rate of 2.5 $\mu g/m^2/day$.

Once SO_2 and H_2S are in aerosol form as SO_4, precipitation scavenging by clouds and rain are effective removal processes. Particles are also removed from the atmosphere as dry fallout. For precipitation processes the rate of scavenging is dependent upon intensity of precipitation activity and on the size of the aerosol.

C. SO_2 BACKGROUND CONCENTRATIONS

Sulfur as SO_2 and SO_4 has been measured in polluted atmospheres for many years, and voluminous statistics are available. However, in our analysis of the total cycle of sulfur in the environment concentration data are needed for the clean ambient atmosphere. These data are very sparse.

Vertical profiles taken over Nebraska indicate values of less than 0.3 ppb SO_2 in the upper portions of the troposphere (Georgii, 1967). These are in line with the few other available measurements of surface SO_2 in very remote places, such as the 0.3 ppb found in Hawaii and 1 ppb on the southeast coast of Florida by Junge (1963). Similar values, of 0.3–1 ppb SO_2, were recently found by Cadle *et al.* (1968) in Antarctica and by Lodge and Pate (1966) in the Panama Canal Zone. Over wide areas of the Central Atlantic, Kühme (1967) found no SO_2 above the limit of detection, which was about 0.3 ppb. From these values we have tentatively concluded that the average tropospheric SO_2 concentration on a global basis is about 0.2 ppb.

Table II summarizes our present best estimates of background concentrations for SO_2 and the other important atmospheric sulfur compounds.

D. THE ENVIRONMENTAL SULFUR CYCLE

Our present calculations, along with additional data from the literature, permit us to estimate the circulation of sulfur in various compound forms through our environment. Figure 1 shows our estimate of this sulfur circulation. Some of the values used in this calculation are reasonably well known, i.e., pollutant emissions and total depositions, but some data must be considered very speculative and have been adjusted reasonably to balance the cycle, e.g., land and sea emissions of H_2S.

TABLE II
Average tropospheric concentrations of sulfur compounds

Compound	Average concentration	Average concentration as sulfur
SO_2	0.2 ppb	0.25 $\mu g/m^3$
H_2S	0.2 ppb	0.14 $\mu g/m^3$
SO_4	2 $\mu g/m^3$	0.7 $\mu g/m^3$

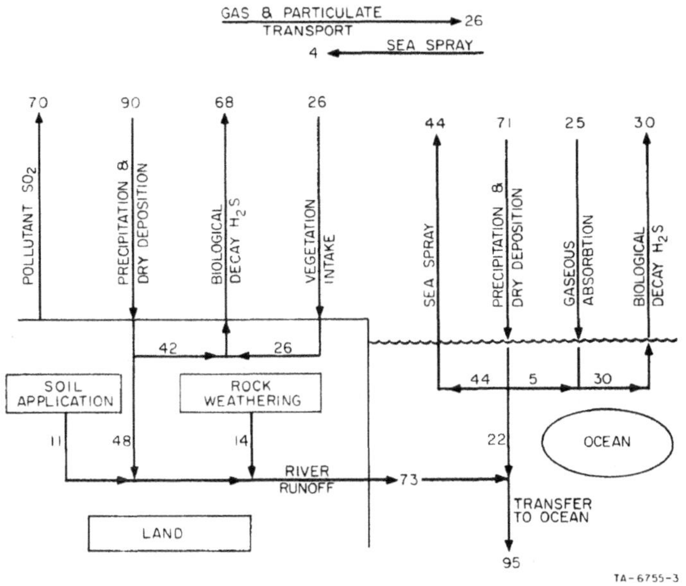

Fig. 1. Environmental sulfur circulation. Units: 10^6 tons/yr sulfur.

To understand this circulation better, we can examine its various components. The sulfur annually discharged to the sea by the world's rivers is 73×10^6 tons; this results from sulfur accumulated from weathering rocks, 14×10^6 tons; sulfur applied to the soil as fertilizer, etc., 11×10^6 tons; and sulfur deposited on the soil by precipitation of dry deposition, 48×10^6 tons. These amounts were estimated by Eriksson (1960).

The atmosphere-land portion of the cycle contains 70×10^6 tons of pollutant sulfur as SO_2 and H_2S emitted to the atmosphere; 90×10^6 tons of sulfur, mostly as SO_4, deposited from the atmosphere to the land; a loss of 68×10^6 tons of sulfur as H_2S from decaying vegetation, and an intake of sulfur by vegetation from the atmosphere of 26×10^6 tons. The 90×10^6 tons deposited includes 80%, or 70×10^6 tons, in rain and the remainder as dry deposition (Junge, 1963). As indicated, 48×10^6 tons of this deposited sulfur is carried off by rivers, and 42×10^6 tons is absorbed by vegetation and then released as H_2S. The intake of sulfur by vegetation is estimated to be 26×10^6 tons based on calculations using a deposition velocity of 1 cm/sec and an ambient concentration of 0.4 ppb (0.5 $\mu g/m^3$). This is twice the average tropospheric concentrations listed in Table II, which are based on the argument that ground level concentrations over land would be higher than the average for the whole troposphere. The 68×10^6 tons estimated for the emission of sulfur as H_2S from vegetation decay results from a summation of the atmospheric vegetation intake, 26×10^6 tons, and the excess of deposition over river carryoff, 42×10^6 tons. This assumes that there is no net accumulation in the surface soils, which seems reasonable. This value is close to Eriksson's (1960) estimate of 77×10^6 tons for H_2S emissions from land areas; however, no data are available with which to check this value. Land areas also gain 4×10^6 tons of sulfur from sea spray (Eriksson, 1959).

The net result of this land circulation is an excess of 26×10^6 tons of sulfur which must be deposited in the ocean if there is to be no net accumulation in the atmosphere.

Deposition of sulfate in the ocean in rain and as dust is 71×10^6 tons (Junge, 1963). The ocean also absorbs 25×10^6 tons of gaseous sulfur calculated on the basis of an SO_2 concentration of 0.2 ppb and a deposition velocity of 0.9 cm/sec (Eriksson, 1959, 1960). The ocean surface is a source of 44×10^6 tons of sulfur in sea spray (Eriksson, 1960) and 30×10^6 tons of sulfur as H_2S from vegetation decay. There is a tropospheric transfer of 4×10^6 tons of sulfur from the ocean to the land. The H_2S emission of 30×10^6 tons is obtained on the basis of what is needed to balance the 100×10^6 tons of gaseous and solid pickup by the ocean and the transfer from sea to land. There are no data that would provide a check as to whether or not this is reasonable. It is a significantly smaller value than the approximately 200×10^6 tons estimated by Eriksson (1960) and Junge (1963) for similar calculations.

The end result of this cycle is an accumulation of sulfur in the oceans of 95×10^6 tons, which is the sum of pollutant emissions, sulfur applied to the soil, and rock weathering.

This compilation of facts and discussion about sulfur in our environment points up a number of interesting things especially relative to pollutant and natural sources of sulfur. With regard to estimates of gaseous sulfur emissions, our evaluation of available data indicates that natural emissions of sulfur, in the form of H_2S, are about 30% greater than are the estimated industrial emissions of SO_2 and H_2S, i.e., 100×10^6 tons as sulfur from H_2S, compared to 76×10^6 tons as sulfur from SO_2. With regard to sulfur pollutants, the most significant fact is that SO_2 is the only significant pollutant and the transformation of SO_2 to SO_4 in the form of H_2SO_4 occurs in a matter of days, perhaps about four. Most of the emitted SO_2 becomes SO_4 in the atmosphere as a result of several possible photochemical or physical reactions. This rapid reaction rate plus ready absorption of SO_2 by vegetation contributes to a rapid decrease in concentration outside emission source areas. In the ambient troposphere the majority of the sulfur is present as SO_4.

It is unlikely that we can adequately evaluate the circulation of sulfur and the relative importance of the various sulfur compound sources until considerably more data are gathered over the oceans and remote land areas of the world.

3. NO_2 and the Nitrogen Cycle

The total circulation of nitrogen compounds through the earth's environment is relatively little known. However, since there are numerous compounds involved, the nitrogen circulation is obviously a complex one. Of the seven significant atmospheric forms of nitrogen, the only significant pollutants emitted by man's activities are NO and NO_2.

A. SOURCES OF NO_2

There are indications that significant amounts of NO_2 can result from NO formed by bacterial action in the reduction of nitrogen compounds under anaerobic conditions. For example, Junge (1963) points out that the formation of NO from HNO_3 occurs in acid soils.

The fixation of nitrogen by lightning has been investigated in a number of studies, but most writers conclude that lightning is unimportant in the fixation of nitrogen (e.g., Georgii, 1963). It also seems unlikely that stratospheric processes are significant in the lower atmospheric nitrogen cycle.

The pollution sources of NO and NO_2, usually considered together and expressed as NO_2, are primarily those combustion processes in which the temperatures are high enough to fix the nitrogen in the air and in which the combustion gases are quenched rapidly enough to reduce the subsequent decomposition. Worldwide No_2 emissions can be roughly estimated on the basis of estimates of fuel combustion processes using available NO_2 production ratios (Mayer, 1965).

Table III summarizes our estimates of the emission sources of NO_2 and other important nitrogen compounds. The data on pollutant emissions are based on average source emissions. The natural emissions are primarily based on the cycle of nitrogen compounds that has been derived in this study and on the amounts that would be needed to balance the cycle. Probably of most interest is the balance between pollutant and natural emissions of NO_2. It is our estimate that these are in the ratio of about 1 to 7, the natural emissions being much greater than the pollutant emissions. Ammonia, a product of biological decay, is the predonimant nitrogen compound emitted to the atmosphere.

TABLE III

Estimated annual global emissions of nitrogen compounds

Compound	Source	Source magnitude (tons/yr)	Estimated emissions (tons/yr)	Emissions as nitrogen (tons/yr)
NO_2	Coal combustion	3074×10^6	26.9×10^6	8.2×10^6
	Petroleum refining	$11\,317 \times 10^6$ (bbl)	0.7×10^6	0.2×10^6
	Gasoline combustion	379×10^6	7.5×10^6	2.3×10^6
	Other oil combustion	894×10^6	14.1×10^6	4.3×10^6
	Natural gas combustion	20.56×10^{12} (ft^3)	2.1×10^6	0.6×10^6
	Other combustion	1290×10^6	1.6×10^6	0.5×10^6
Total NO_2			52.9×10^6	16.1×10^6
NH_3	Combustion		4.2×10^6	3.5×10^6
NO_2	Biological action		500×10^6	150×10^6
NH_3	Biological action		5900×10^6	4900×10^6
N_2O	Biological action		650×10^6	410×10^6

B. ATMOSPHERIC NO_2 CONCENTRATIONS

Data on NO_2 concentrations outside city areas are widely scattered, but show definitely that NO_2 is present in trace amounts in remote locations. Analyses by Lodge and Pate (1966) in Panama indicate a dry season average for NO_2 of 0.9 ppb and a rainy season value of 3.6 ppb. A few samples of NO have indicated concentrations up to 6 ppb. Junge (1956) has reported NO_2 concentrations in Florida averaging 0.9 ppb and Hawaiian data averaging about 1.3 ppb.

In Ireland O'Connor (1962) reported average NO_2 concentrations for North Atlantic air to be 0.3 ppb in April and 0.2 ppb in September and October. More recently NO_2 and NO data from Pike's Peak, Colorado, and from a remote area of North Carolina have been reported (Hamilton *et al.*, 1968; Ripperton *et al.*, 1968). The Colorado data indicate an average NO_2 concentration of 4.1 ppb and 2.7 ppb for NO. In the Appalachian area of North Carolina, average concentrations were 4.6 ppb for NO_2.

From these scattered data we have concluded that in continental areas average levels of NO_2 are about 4 ppb. Ocean areas appear to be characterized by lower concentrations, perhaps 1.0 ppb, or less.

Table IV summarizes our estimates of atmospheric concentrations of nitrogen compounds, and the mass of nitrogen present in these compounds throughout the atmosphere. In this tabulation N_2O accounts for more than 97% of the tonnage, calculated on the basis of N content. Of the 3% of the materials which show a certain degree of reactivity and change, 80% of the N is present as NH_3.

C. ATMOSPHERIC SCAVENGING REACTIONS FOR NO_2

NO and NO_2 are important as pollutants mainly because of their participation in photochemical reactions, which constitute a major scavenging process.

TABLE IV
Background concentrations of nitrogen compounds

Compound	Ambient concentration	Atmospheric mass as *N*
N_2O	0.25 ppm	1500×10^6 tons
NO/NO_2		
Land	4 ppb (as NO_2)	12×10^6
Ocean	1 ppb	3×10^6
NH_3	6 ppb	30×10^6
NO_3	0.2 $\mu g/m^3$	0.2×10^6
NH_4	1.0 $\mu g/m^3$	4.1×10^6

The reaction of NO with O_3 in the atmosphere to form NO_2 is rapid, and since there is always some background O_3 from stratospheric transport, it is generally assumed that NO_2 rather than NO is the predominant of the two species found in clean atmospheres. However, the previously quoted Panama data of Lodge and Pate (1966) and the Colorada data of Hamilton *et al.* (1968) indicate that this assumption for the ambient atmosphere may have to be altered.

The O_3 oxidation of NO_2 to N_2O_5 is about 500 times slower than the O_3 oxidation of NO; nevertheless it is rapid enough to limit the half-life of 1 ppb NO_2 in the atmosphere in the presence of 5 ppb of O_3 to about two weeks. However, the residence time of NO_2 based on our atmospheric nitrogen cycle is only three days. This short time can probably be explained by the following scavenging reaction of NO_2:

$$3NO_{2(v)} + H_2O_{(v)} \leftrightarrows 2HNO_{3(v)} + NO_{(v)}$$

The equilibrium constant is 0.004 atm^{-1} (McHenry, 1953) at 25°C, but with the extreme excess of water vapor found in the atmosphere 10% of the NO_2 is converted to HNO_3 at equilibrium. This HNO_3 vapor is rapidly removed by reaction with atmospheric ammonia and absorption into hygroscopic particles. At relative humidities higher than 98%, condensation of dilute nitric acid droplets will occur at 10°C and a HNO_3 vapor concentration of 0.1 ppb. All of the HNO_3 eventually becomes nitrate salt aerosol.

D. NO_2 AND THE GENERAL NITROGEN CYCLE

This analysis of atmospheric nitrogen compounds provides us with some rough estimates of global sources and sinks. We have combined these source and sink data into a simplified nitrogen circulation as shown in Figure 2. The basic features of this circulation are as follows:

(a) The N_2O cycle is independent of the rest of the system, except that it draws on soil nitrogen for formation. One sink for N_2O is above 30 km and results in the production of 5×10^6 tons per year of NO.

(b) Deposition and destruction of N_2O by soil bacteria is the major sink.

(c) Nitrogen appears to enter the soil regime through fixation in biological processes. Hutchinson (1954) calculated an annual nitrogen fixation rate of 130×10^6 tons.

(d) NH_4 particulate deposition, 3500×10^6 tons, is determined from NH_4 content of rainfall, 2800×10^6, plus dry deposition of 25%. The source of the NH_4 is gaseous NH_3 released from the biosphere.

(e) NH_3 gaseous deposition, 900×10^6 tons, was obtained by analogy with SO_2 using a deposition velocity, v_g, of 1 cm/sec and a concentration of 6 ppb.

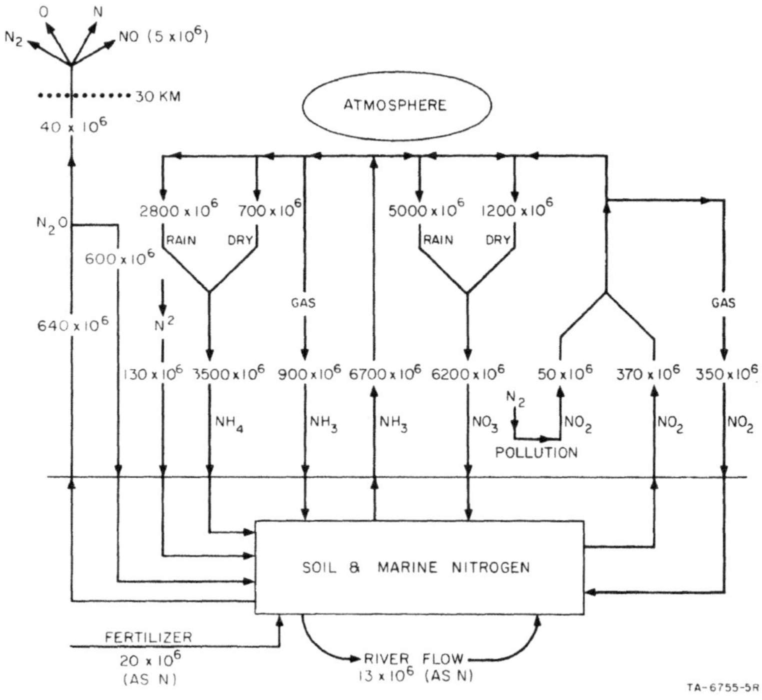

Fig. 2. Simplified nitrogen compound circulation. Units: tons/yr nitrogen.

(f) NH_3 gaseous emissions were calculated to balance gaseous NH_3 deposition and particulate depositions of both NH_4 and NO_3. This resulting emission rate, 5900×10^6 tons, indicates a residence time of about two days.

(g) NO_3 particulate deposition, 6200×10^6 tons, is determined by the NO_3 content of rainfall, 5000×10^6 tons, plus 25% for dry deposition. A major source for the nitrogen in this NO_3 deposition is NH_3 from the biosphere.

(h) NO_2 pollution emissions are 50×10^6 tons. The source of this nitrogen is the atmosphere.

(i) NO_2 from the biosphere, 370×10^6 tons, provides sufficient NO_2 to balance an atmospheric residence time of about five days.

(j) NO_2 gaseous deposition was obtained using a deposition velocity, v_g, of 0.5 cm/sec and a global average concentration of 2 ppb.

(k) Tonnages of nitrogen added to the soil annually as fertilizer, 20×10^6 tons (U.S. Statistical Abstracts, 1967), and the nitrogen carried to the ocean by rivers, 13×10^6 tons (Hutchinson, 1954), are included.

This circulation is very much simplified; e.g., no division is made for conditions over the land and the ocean. However, various parameters which result from this initial attempt seem to be of the right order of magnitude, and thus perhaps it is useful to draw one or two conclusions from it.

First, it appears that the dominant sub-cycle is the circulation of NH_3, including its deposition as both gaseous and particulate material. Second, and perhaps more important, the most reasonable way to balance the observed NO_3 deposition is through some reaction whereby NH_3 is oxidized to NO_3. To account for the large amount of NO_3 deposition by increasing the emission of NO_2 would result in a much shorter NO_2 residence time if our global average NO_2 concentration of 2 ppb is generally applicable. It is obvious that much needs to be done before we can fully appreciate the environmental circulation of nitrogen compounds.

This cycle does show even in this simplified form that the pollutant emissions of NO and NO_2 are relatively unimportant factors in the total circulation of nitrogen through the atmosphere.

4. Atmospheric Carbon Monoxide

Carbon monoxide has been considered an important toxic atmospheric pollutant for many years because of its prevalence in automobile exhaust and in the effluents from combustion.

A. SOURCES OF CARBON MONOXIDE

The only significant sources of CO which can be definitely identified on a global scale are pollutant sources. The major combustion source is the gasoline engine. Lesser amounts of CO are also produced by coal combustion, industrial operations, boilers, furnaces, and waste disposal. Some CO is also produced by forest and prairie fires, and minor amounts are formed by photochemical hydrocarbon reactions. Estimated emissions of CO on a worldwide basis total 257×10^6 tons. The estimated amounts from various sources are given in Table V. It is estimated that 95% of this CO is produced in the northern hemisphere.

Until very recently natural sources of CO have been completely discounted, but new research now seems to indicate that natural sources may exist in both the ocean and over land areas.

The first indication of possible marine involvement in the CO cycle was found on a trans-Pacific oceanographic cruise of the USNS Eltanin from San Francisco to New Zealand in the fall of 1967. On this cruise there were frequent diurnal variations in CO concentration which could be logically explained if the surface ocean waters were acting as a source of CO (Robinson and Robbins, 1968b). Variations in CO concentrations between the northern and southern hemisphere also occurred, but these were not comparable to the variations in pollution sources between the two hemispheres.

TABLE V
Estimated global emissions of CO

Source	Consumption (tons/yr)	CO emission (tons/yr)
Gasoline	379×10^6	193×10^6
Coal	3074×10^6	12×10^6
Wood (fuel)	466×10^6	16.0×10^6
Incineration	500×10^6	25.0×10^6
Forest fires	18.0×10^6*	11.3×10^6
	Total	257×10^6

* Acres per year.

More recently, Swinnerton *et al.* (1969) at the Naval Research Laboratory in Washington, D.C., have made comparative CO measurements in both the atmosphere and the ocean in the Caribbean area. Their measurements show that the surface water in the open ocean was supersaturated relative to atmospheric CO concentrations. On a quantitative basis this supersaturation in one area was as much as 90 times and, in general, the ocean CO concentration was at least 10 times the equilibrium saturation. Atmospheric concentrations over the open ocean ranged from about 75 to 250 ppb, which is in the same range as that encountered in the Pacific. These NRL data can best be explained by postulating a source of CO in the surface ocean water. Although the source of this CO in the ocean has not been found, a logical explanation is that it is a biological process.

Another potential natural source of CO over land areas follows from Went's proposal that photoreactions of terpenes are significant in the atmosphere (Went, 1966). He estimates that 10^9 tons of volatile organics of plant origin are molecularly dispersed throughout the world each year. Photochemical oxidation of these plant volatiles by ozone or nitrogen oxide could produce 61×10^6 tons of CO annually. We are not, at this time, considering possible direct CO production by vegetation as described by Wilks (1959).

B. ATMOSPHERIC CO CONCENTRATIONS

The presence of CO in urban atmospheres at concentrations as high as 50–100 ppm has been recognized for many years, but the determination that CO was present in trace amounts in the ambient atmosphere did not occur until 1949 when Migeotte (1949) detected absorption lines in the solar spectrum around 4.7 μ and attributed them to the presence of CO in the earth's atmosphere.

In the past few years CO concentration data from remote locations have become available and it is now possible to form a tentative model of tropospheric CO concentrations. Data from five ship crossings of the Pacific indicate quite variable concentrations in the northern hemisphere, the average being between 0.1 and 0.2 ppm (Robinson and Robbins, 1968). In the Arctic, sampling indicates a background level of about 0.1 ppm (Cavanagh *et al.*, 1969; Robinson and Robbins, 1969). In the southern hemisphere concentrations still show considerable variation, but judging from Pacific data they are apparently much lower, averaging about 0.06 ppm. Data taken by Junge (1968) at various altitudes over Europe also indicate a northern hemispheric CO background of between 0.1 and 0.2 ppm.

For our tentative model we have used 0.20 ppm for a northern hemisphere average concentration and 0.06 for a southern hemisphere value. The 0.2 ppm value is somewhat higher than the north Pacific average in order to account for higher concentrations that are expected to be characteristic of most continental areas.

C. SCAVENGING PROCESSES FOR CO

Three theoretical sink mechanisms exist – chemical, physical, and biological. Under any reasonable atmospheric conditions of concentration and temperature, photochemical processes are the only chemical reactions that can result in the oxidation of CO to CO_2. However, these can only occur in the high atmosphere, above 100 km, and they certainly are of very minor importance in the total atmospheric CO cycle. No direct reaction at the earth's surface to produce a carbon compound other than CO can be postulated.

Physical adsorption onto various surfaces cannot be substantiated by referring to any known parameters of CO. The possibility of CO being absorbed in the oceans has been examined, and it can be shown to be an insignificant sink, since there is no known reaction between CO and other chemical or biological constituents of seawater.

The possibility that a biological sink for CO might exist on land does not seem as unfavorable as do the cases of the chemical or physical processes. Experiments have shown that certain bacteria will reduce CO and produce methane. However, it must be admitted that there is no available experimental evidence to define any metabolic action between CO and plants. It is known, however, that plants are effective scavengers for a wide variety of atmospheric material, including CO_2, SO_2, O_3, H_2S, and fluorides. Furthermore, many chemical reactions can occur within a plant, and it is not unreasonable to suppose that CO could be involved in some of them. Even without a specific CO reaction process, however, absorption within the plant either on surfaces or in fluids could provide a means of tying up a large amount of CO; if the process were not readily reversible, this could be a significant sink. It also might be possible for some biological processes in the ocean to provide a means of tying up CO.

At present we believe that a biological process is the CO sink which seems to be required to balance the emission and atmospheric data. This supposition is not based on any available research data, but rather on the fact that there are apparently no data to show that a biological process could not be the CO sink. However, considerable research is being carried out on this question at the present time.

D. AN ATMOSPHERIC CO CYCLE

With the absence of any well-defined scavenging mechanism, it is not possible to postulate even a crude environmental cycle for CO. However, some useful related material can be derived about the circulation of CO.

One factor is the average atmospheric residence time, τ, which is the ratio of the CO in the atmosphere to the total annual emission. Considering the average global CO concentration, 0.13 ppm, and the global pollutant emission, 257×10^6 tons/year, gives a calculated residence time of 3.1 years. If the emission rate is increased by the amount of CO resulting from the proposed photochemical reactions of naturally emitted terpenes, 61×10^6 tons/year, the residence time is reduced to 2.5 years. When an additional CO source in the ocean is also considered, a residence time for CO of about two years seems most likely.

To maintain an environmental circulation of 2–3 years would require a reasonably effective scavenging mechanism; but as we previously mentioned, none seems apparent in the environment. The differences in northern and southern hemispheric CO concentrations seem to be most logically attributable to the presence of 95% of the pollution sources in the northern hemisphere.

These two aspects of our crude CO cycle are obviously worthy of considerable further research.

5. Summary

This analysis of pollutant emissions has attempted to relate the gaseous emanations from both man-made and natural sources to the circulation through the atmosphere of several common pollutants. In some instances natural sources on a global basis are of greater magnitude than are the urban pollution sources. Some of the most important of the natural sources are in the nitrogen cycle, where emissions of both NO_2 and NH_3 are predominantly from natural sources in the biosphere. In the circulation of sulfur compounds, the emission of H_2S from biological processes is estimated to be a major source of sulfur which eventually is detected as a sulfate aerosol. In the CO cycle there are strong indications that natural sources are present – both in the biosphere and in the ocean.

The fact that trace amounts of a wide variety of compounds emanate from the natural environment is important for several reasons. The presence of these natural sources emphasizes the fact that there are processes present in the atmosphere which can scavenge the common pollutants emitted from urban sources. In the case of the more reactive gases – SO_2, NO_2, etc. – these scavenging processes act relatively rapidly. The scavenging cycle is apparently as short as a day or so for SO_2 and as long as a few years for CO.

These scavenging cycles have to be recognized if we are going to achieve an understanding of our environment. However, the fact that these cycles exist cannot be used as an excuse for permitting any urban area to reach levels of pollution that are detrimental in any way to its residents. None of the scavenging cycles that have been described are effective enough within the time-frame of a few hours for them to have any significant effect within a given source area. Thus, even with these natural processes some form of control of pollutant sources must be used to protect a local area. However, we can apparently rest somewhat more easily with regard to the possibility of the accumulation of many pollutants on a global scale. The natural scavenging processes are effective when the time scale is increased.

There are many aspects of the cycling of trace materials through the atmosphere that we do not understand. As our understanding increases we can more adequately use the atmosphere as a renewable resource and also conserve its properties as one of our most important environmental factors.

Acknowledgments

This study is part of a larger research program on gaseous contaminants in the atmosphere supported by the American Petroleum Institute. The authors are pleased to acknowledge this support.

References

Bielke, S. and Georgii, H.-W.: 1967, *Tellus* **20**, 435–41.
Cadle, R. D., Fischer, W. H., Frank, E. R., and Lodge, J. P.: 1968, *J. Atm. Sci.* **25**, 100.
Cavanagh, L. A., Schadt, C. F., and Robinson, E.: 1969, *Environmental Science and Technology* (in press).
Chamberlain, A. C.: 1960, *Int. J. Air. Poll.* **3**, 63.
Eriksson, E.: 1959, Part I, *Tellus* **11**, 375–403.
Eriksson, E.: 1960, Part II, *Tellus* **12**, 63–109.
Georgii, H.-W.: 1963, *J. Geophys. Res.* **68**, 3963.
Georgii, H.-W.: 1967, personal communication to C. E. Junge.
Hamilton, H. L., Worth, J. J. B., and Ripperton, L. A.: 1968, An Atmospheric Physics and Chemistry Study on Pikes Peak in Support of Pulmonary Edema Research, Research Triangle Institute, North Carolina, for Army Research Office, Contract No. DA-HC19-67-C-0029.
Hutchinson, G. E.: 1954, in *The Earth as a Planet* (ed. by G. P. Kuiper), University of Chicago Press, Chicago.
Junge, C. E.: 1956 *Tellus* **8**, 127.
Junge, C. E.: 1963, *Air Chemistry and Radioactivity*, Academic Press, New York, p. 123.
Junge, C. E.: 1968, private communication.
Junge, C. E. and Ryan, T.: 1958, *Quart. J. Roy. Meteorol. Soc.* **84**, 46–55.
Katz, M.: 1958, in *Air Pollution Handbook* (ed. by P. L. Magill, F. R. Holden, and C. Ackley), McGraw-Hill Book Company, Inc., New York.
Katz, M. and Ledingham, G. A.: 1939, National Research Council of Canada, in *Effect of Sulfur Dioxide on Vegetation*, NCR No. 815, Ottawa.
Kühme, H.: 1967, private communication to C. E. Junge.
Lodge, J. P. and Pate, J. B.: 1966, *Science* **153**, 408.
Mayer, M.: 1965, 'A Compilation of Air Pollution Emission Factor'. U.S. Public Health Service, Division of Air Pollution, Cincinnati, Ohio
McHenry, L. R. J.: 1953, 'The Vapor Phase Reaction between Nitrogen Oxides and Water'. M.S. thesis in chemical engineering, University of Illinois.
Migeotte, M. V.: 1949, *Phys. Rev.* **75**, 1108.
O'Connor, T. C.: 1962, 'Atmospheric Condensation Nuclei and Trace Gases'. Final Report, Dept. of Physics, University College, Galway, Ireland, Contract No. DA-91-591-EUC-2126.
Renzetti, N. A. and Doyle, G. J.: 1960, *Int. J. Air Poll.* **2**, 327.
Ripperton, L. A., Worth, J. J. B., and Kornreich, L.: 1968, 'Nitrogen Dioxide and Nitric Oxide in Non-Urban Air'. Paper No. 68-122, 61st Annual Meeting Air Pollution Control Association, June.
Robinson, E. and Robbins, R. C.: 1968a, 'Sources, Abundance, and Fate of Gaseous Atmospheric Pollutants'. Final Report SRI Project PR-6755, for American Petroleum Institute, New York, February.
Robinson, E. and Robbins, R. C.: 1968b, *Antarctic Journal of the U.S.* **194**, Sept-Oct.
Robinson, E. and Robbins, R. C.: 1969, *J. Geophys. Res.* (in press).

Swinnerton, J. W., Linnenbom, V. J., and Check, C. H.: 1969, *Envir. Sci. Techn.* **3**, 836.
U.S. Statistical Abstracts: 1967, U.S. Govt. Printing Office.
Went, F. W.: 1966, *Tellus* **18:203**, 549–556.
Wilks, S. S.: 1959, *Science* **129**, 964–966.

Suggested Further Reading

1. C. E. Junge, *Air Chemistry and Radioactivity*, Academic Press, New York, 1963.
2. P. A. Leighton, *Photochemistry of Air Pollution*, Academic Press, New York, 1961.
3. A. C. Stern (ed.), *Air Pollution*, Second edition, Academic Press, New York, 1968, Vol. I, Chapters 2, 6, and 11.

PART II

NITROGEN COMPOUNDS IN SOIL, WATER, ATMOSPHERE AND PRECIPITATION

INTRODUCTION

One of the vexing problems is the conflict between the demands for more food and the demands for a clean environment. Modern, scientific and highly-productive agriculture requires fertilizers, principally nitrogen and phosphorus. It is very susceptible to damage by various pests. It usually requires irrigation, which often involves a transfer of water, and a return flow which carries off nutrients, chlorides, pesticides – all undesirable substances – into streams and rivers.

The addition of fertilizers and pesticides also affects the character of the soil. Over-application especially of nitrates may lead to harmful effects on cattle and produce risks to human health. These and other problems are discussed in the papers by Commoner and Byerly. Keeney and Gardner deal particularly with nitrogen transformations in the soil.

The effects of pollution, whether produced by agriculture, by industry, or by municipal sewage, are most noticeable in the case of lakes, the subject of Hasler's paper. Examples exist throughout the world; it may be appropriate here to detail the case of Lake Erie where an extensive study has just been completed and where action is planned.

Lake Erie, one of our largest lakes, the world's twelfth largest, provides a resource to 11.5 million people in the U.S.A. and Canada in terms of water supply, recreation, commercial fishing and shipping. In addition, the annual value added by manufacturing in the Erie Basin stands at more than $17 billion. By the year 2000, the population will have doubled and so will the industry of the Basin. These people and industries depend on Lake Erie, which forms a priceless national heritage whose quality must be maintained and enhanced, so that it can be passed on to future generations in a condition of unlimited usefulness.

Lake Erie constitutes a warning to mankind, as Professor Barry Commoner points out. Other lakes, both large and small, are experiencing or will soon experience the same fate, as will many of the Nation's important estuaries. Referred to technically as eutrophication, this aging of lakes is accelerated by an oversupply of nutrients which are generated by human activities. It can bring about the premature death of the lake, making it less and less useful, until it finally becomes a stinking nuisance.

Our strategy is twofold: prevention and restoration. Prevention means greatly slowing down eutrophication by removing key nutrients from the wastewater before it enters the lake. Restoration means removing or inactivating the nutrients after they have reached the lake. Restoration techniques have to be carefully researched in order to find an economically acceptable method likely to succeed. Examples are: mechanical harvesting of algae, or harvesting with organisms which eat algae (preferably edible fish), tying up the nutrients by chemical methods (the Swedes are treating lakes experimentally with pelletized aluminum sulphate), flushing with water of low nutrient content, or bringing oxygen to the deeper layers by mechanical or thermal mixing. All of these techniques, and many others, are being studied intensively by the National Eutrophication Research Program of the Federal Water Pollution Control Administration of the Department of the Interior. The work is done in government laboratories, university laboratories, and by private industry and, in many cases, uses small lakes in different parts of the country as field laboratories.

Restoration methods presently proposed are not only expensive but may not be able to do the whole job. Prevention, on the other hand, is less costly, technically simpler and, on the basis of our present knowledge, will improve the condition of the lake – without restoration. Therefore, the FWPCA study, released in August 1968, pinpoints a specific approach; namely, the removal of phosphorus – a key nutrient for the growth of algae – from the wastewater released by municipal treatment plants. Nitrogen is also a key nutrient for algal growth. However, its removal is more difficult and, in any event, it can enter the lake in rainfall or be captured from the atmosphere by certain blue-green algae that can fix nitrogen.

Based on past research, we know that chemicals can precipitate phosphates out of waste water, and do the job quite economically. For example, FWPCA figures show that phosphate precipitation by lime would add a cost of less than 7 cents per thousand gallons in a 10 million gallon per day plant, or, roughly, an increase of 34%.

Fortunately, most major sources of phosphorus are 'point sources'; i.e., the wastewater comes out of a pipe and can be treated. About $\frac{3}{4}$ of the load comes from municipal wastes, which contribute about 3 pounds per person per year; $\frac{2}{3}$ of this comes from detergents. (Interestingly enough, a domestic duck produces nearly 1 pound per year.) Certain industries, such as potato processing, add 1.7 pounds per ton. As a matter of high priority, the Soap and Detergent Association, together with the Department of the Interior are exploring ways and means of doing something about the phosphates in detergents. It is more difficult to control non-point sources; for example, agricultural drainage, unless appropriate land management measures are practised, especially proper methods of applying fertilizers.

We have the key tools: the basic laws now on the books, the technical competence, and the equipment. Together with adequate funds, these would do the job. During the past year, water quality standards have been approved for all of the Great Lakes States, under the terms of the Federal Water Pollution Control Act. Based on existing technology, the cost of treatment facilities over the next five years, sufficient to meet the requirements for municipal and industrial wastes, is estimated to be approximately $1.4 billion. A fraction of this money is expected to come from federal funds; the remainder has to be provided by municipalities and by industries. Most of the research is done under federal auspices, but the amounts specifically earmarked for eutrophication are small, only $1.8 million per year; yet on this research depends the effectiveness of a major national program.

Some recommend more extreme measures and would make the restoration of Lake Erie a national project, similar in scope and size to the space program. It could involve tremendous engineering projects including even the physical removal of the polluted sediments. But before we undertake such tasks and commit ourselves to vast expenditures involving many billions of dollars, we must take immediate steps which we think have a good chance of success and which cost less.

It is rather unlikely that Lake Erie can feasibly be returned to the condition which existed prior to man's appearance, or even to the condition which existed at the turn of the century. It can return to some intermediate stage of aging, but the exact stage cannot be predicted. Most important, we can expect a major improvement and protection of water quality.

THREATS TO THE INTEGRITY OF THE NITROGEN CYCLE: NITROGEN COMPOUNDS IN SOIL, WATER, ATMOSPHERE AND PRECIPITATION

BARRY COMMONER

Dept. of Botany and Center for the Biology of Natural Systems, Washington University, St. Louis, Mo., U.S.A.

Abstract. Intensified application of nitrogenous fertilizers in the United States has produced severe stress of the natural nitrogen cycle by introducing unprecedented amounts of inorganic nitrogen. Thus, whereas the annual nutritional turnover of nitrogen amounts to 7 or 8 million tons, technology and agriculture introduce into the cycle a nearly equivalent amount of 10 million tons of nitrogen compounds. Soil bacteria rapidly convert nitrogenous fertilizer to nitrates, a large proportion of which drain off into streams and rivers. The result is increased eutrophication and potential hazard from nitrite poisoning, or methemoglobinemia. Surveys show that: (1) nitrate acquired by the Missouri River from Nebraska farmlands has increased in parallel with the increasing annual use of nitrogenous fertilizer in Nebraska since 1955 and that (2) high nitrate content of rivers in Illinois farmland is traceable almost entirely to fertilizer that drains into groundwater. Increased use of inorganic N fertilizer also appears to be associated with certain instances of dangerously high nitrate levels in vegetable foods. In the U.S. nitrogen oxides from automobile exhausts also represent an important stress on the N cycle.

1. Introduction

Four chemical elements make up the bulk of living matter – carbon, hydrogen, oxygen, and nitrogen – and they move in great, interwoven cycles in the surface layers of the earth: now a component of the air or water, now a constituent of a living organism, now part of some waste product, after a time perhaps built into mineral deposits or fossil remains. Nitrogen occupies a special place among these four elements of life because it is so sensitive an indicator of the quality of life. Nitrogen deprivation is a first sign of human poverty; a certain outcome is poor health, for so much of the body's vital machinery is made of nitrogen-bearing molecules: proteins, nucleic acids, enzymes, vitamins, and hormones. Nitrogen is, therefore, closely coupled to human needs. Indeed, as the world population grows, nitrogen will become, increasingly, the crucial element in our efforts to avert catastrophic famine.

All living things, by their very life processes, can alter the precise balance of the natural cycles. In nature these processes affect the chemistry of the environment slowly, on the time scale of geological events, and the systems of living organisms have time to adjust to them. Only one living thing, man, has the power – through technology – to make changes that are so rapid, and so large compared with the natural cycles as to stress them to the point of collapse.

The nitrogen cycle is very vulnerable to human intervention. In the U.S., the overall annual nutritional turnover of nitrogen amounts to about 7 or 8 million tons. At the present time technology and agriculture introduces into this cycle about 7 million tons of nitrogen compounds from chemical fertilizers, and about 2–3 million tons of nitrogen compounds generated by automobile exhausts and power plants.

The purpose of this paper is to discuss what we know and need to know about the consequences of this intrusion into the natural nitrogen cycle. The evidence will, I believe, support the conclusion that modern technology has intruded into the cycle at its most vulnerable point. This intrusion has produced stresses which have already contributed to a serious deterioration of the quality of the environment. If continued they threaten to collapse important segments of the cycle itself. Correction of these trends will require us to solve very grave economic, social, and political problems.

In the ecosphere – the earth's thin skin of soil, water, air, and living things – nitrogen is found in relatively few basic chemical forms. A striking feature of nitrogen chemistry is that combinations of nitrogen and oxygen are relatively rare. The great bulk of the earth's nitrogen is represented by the nitrogen gas in the air. In living things nitrogen almost invariably occurs in combination with hydrogen rather than oxygen (i.e., the nitrogen compounds tend to be in relatively reduced rather than oxidized states); only a few natural biochemical substances, among the many thousands known, contain nitrogen in combination with oxygen. In nature there are only relatively small amounts of the molecular forms – nitrate, nitrite, and several gaseous nitrogen oxides – in which nitrogen and oxygen are combined. Thus, we live in an environment in which nitrogen, as it passes through the successive stages of its cycle rarely finds itself, for long, (except in inorganic mineral deposits) combined with oxygen.

It is precisely this slender span of the nitrogen cycle which is most affected by human intrusion, for the nitrogen introduced into the environment by technology is almost entirely in oxidized forms. Regardless of its original chemical state, nitrogen fertilizer is rapidly converted to nitrate by soil bacteria; automobile exhausts and power plants generate various oxides of nitrogen. These intrusions upon the natural nitrogen cycle have developed very rapidly. In the United States, during the last 25 years the annual consumption of inorganic nitrogen fertilizer has increased about 14-fold and automotive activity has increased about three-fold. In this period, *biological* intrusion on the nitrogen cycle due to human activity has risen more slowly; nitrogen in sewage has increased about 70%.

The maintenance of the naturally low concentrations of oxidized forms of nitrogen is essential to the integrity of the earth's life system. Important hazards to this system are generated when the concentrations of these nitrogen compounds are artificially increased. One hazard is pollution of surface waters by excessive amounts of nitrate. When the normally low level of nitrate in natural waters is increased, the growth of algae may be sharply enhanced. The resulting 'algal blooms', which soon die, overburden the water with organic matter, which on being oxidized by microorganisms depletes the oxygen content of the water, causing the natural cycles of self-purification to collapse. The Spilhaus report estimates that by 1980 the burden of organic matter imposed on surface waters will be sufficient to consume the total oxygen content of the summertime flow of every river system in the U.S. Excessive amounts of nitrate may contribute to this potential collapse of the self-purifying processes of the nation's water systems.

A second hazard is due to the potential toxicity of nitrate to human beings and domestic animals. While nitrate is itself relatively innocuous when ingested, it may be converted by intestinal bacteria into nitrite – which, on combining with hemoglobin in the blood (methemoglobinemia) destroys the oxygen-carrying power of the blood, resulting in asphyxia, and possibly death. Another hazard is that nitrite, on reaction with certain organic constituents of food, or even of the body itself, may form nitroso compounds, some of which are powerful carcinogens. A third hazard of imbalance in the nitrogen cycle is that certain of the gaseous nitrogen oxides are powerful poisons and pharmacological agents. Finally, such oxides, when activated by sunlight, can react with waste hydrocarbons in the air to produce the noxious constituents of smog.

In general, then, we must regard any appreciable elevation in the normally low concentrations of oxidized forms of nitrogen as a serious hazard to the environment. Since it is precisely this most vulnerable segment of the nitrogen cycle which is stressed by recent increases in the nitrogen produced by human activities, it becomes urgent to determine to what extent such stresses are present or potential threats to the quality of the environment.

2. The Natural Nitrogen Cycle

The soil is a useful place to begin. Nitrogen is crucial in the soil economy. In nature, let us say a plant growing in a wood or meadow, most of the soil's nitrogen – usually well over 80% – is in the form of complex and still poorly-understood organic substances, especially humus. Slowly, bacteria release nitrate from humus and decaying organic wastes – manure and the residues of animals and plants. The resultant concentration of nitrate in the soil water is very low and the roots need to work to pull it into the plant. For this work the plant must expend energy which is released by biological oxidation processes in the roots. These processes require oxygen, which can reach the roots only if the soil is sufficiently porous. Soil porosity is governed by its physical structure; in particular a high level of organic nitrogen, in the form of humus, is required to maintain a porous soil structure. Thus, soil porosity, therefore its oxygen content, and hence the efficiency of nutrient absorption, are closely related to the organic nitrogen content of the soil.

When the United States was settled, the soil system was in this natural condition; the soil cycle was in balance, maintaining its nitrogen reserve in the stable organic form. Some nitrogen from the air is fixed by legumes and helps build up the organic reserve. In virgin land only a very small amount of nitrate escapes the plant's root systems and leaches out into surface waters, or escapes to the air in volatile compounds: nitrogen gas, ammonia, and nitrogen oxides; and the last two are soon returned to the earth in rain and snow.

In natural waters a similar nitrogen cycle prevails, except that the large reserve of organic nitrogen represented by the soil humus is lacking. Living things, for example fish, contribute organic nitrogen to the water: waste, and, on death, their bodies. This organic nitrogen is quickly converted to an inorganic form; the bacteria of decay free nitrogen from its organic combination with carbon and hydrogen, and unite it with oxygen to form, ultimately, nitrate. The needed oxygen reaches the water from the air and from the photosynthetic activity of plants. Nitrate in turn nourishes the aquatic plants, chiefly algae; these in turn furnish food for fish and other animals, and the cycle is complete. In a balanced natural system the amounts of organic nitrogen and nitrate dissolved in the water remain low, the population of algae and animals is correspondingly small, and the water is clear and pure. And because the natural nitrogen cycle in the soil is tightly contained, relatively little nitrogen is added to the water in rainfall, or in drainage from the land.

The nitrogen gas that makes up about 80% of the air is very stable, chemically. It will react with oxygen to form nitrogen oxides, and eventually, nitrate, only under intense heat, which in natural circumstances is obtainable only from lightning; however recent evidence suggests that the amounts of nitrogen oxides produced by this natural agency are very slight. In natural conditions animals that breathe air carry to their lungs little else, besides the air's nitrogen, oxygen, water vapor and carbon dioxide. In particular nitrogen oxides, which are incompatible with animals' respiratory physiology, are normally absent.

3. Agricultural Stresses on the Soil

We must introduce into this balanced system, man – not only a rapidly proliferating animal, but also the one living thing that has generated the vast environmental intrusions of technology, agriculture, and urbanization. How has the natural nitrogen cycle been faring under these stresses?

When the United States was settled, the soil system was in the natural condition that I have already described. As this natural system was taken over for agricultural purposes, plants were grown on the soil in amounts much greater than it would sustain in its virgin state. The organic store of nutrients was gradually depleted and crop yields declined year by year. (In the Midwest, the organic nitrogen of the soil is now about one-half of what it was in the virgin soil.) With new lands always available, farmers moved westward repeating the process of skimming from the soil the most available nutrient and leaving it when its productivity fell below a point which made westward migration more attractive. This process came to an end at about 1900 and from then on as agriculture became intensified to meet the demands of a growing population, more and more of the original store of organic nutrients was withdrawn from the soil in the form of crops.

At the turn of the century, then, with the Nation's need for food rapidly rising, and the first easy fruits of the soil's wealth already taken, we were confronted with the vast problem of intensifying the declining agricultural productivity of the soil. Much agricultural research was devoted to sustaining productivity by adding chemical fertilizers to the soil. At several agricultural research stations, such as the Missouri Agricultural Experiment Station, long-term experiments were undertaken to study the effects of different agricultural practices on crop yield, and on the nature of the soil. In 1942, this Station published a remarkably revealing account of 50 years of patient study of their experimental plots, Sanborn Field.

The report showed that nitrogen, provided as nitrate, was an effective means of maintaining good crop yields. But the report also showed that the soil suffered important changes:

The organic matter content and the physical conditions of the soil on the chemically-treated plots have declined rapidly. These altered conditions have prevented sufficient water from percolating into the soil and being stored for drought periods. Apparently a condition has developed in the soil whereby the nutrients applied are not delivered to the plant when needed for optimum growth .... Evidently most of the nitrogen not used by the immediate crop is removed from the soil by leaching or denitrification .... From these figures it is evident that heavy application of chemical fertilizers have given a very low efficiency of recovery. [1]

The story is clear and well supported by numerous laboratory and field investigations: While nitrate fertilizer sustains crop growth, it fails to rebuild the humus nitrogen lost from the virgin soil. As a result soil porosity – which depends on humus content – deteriorates, aeration becomes more difficult and the plant roots, which must have oxygen if they are to absorb nutrients, are unable to take up all of the nitrate made available to them by the added fertilizer. The rest of the nitrate is lost from the soil. Some nitrate leaches out in the ground water, ultimately into rivers and lakes. Some nitrate is converted to ammonia, nitrogen gas, and nitrogen oxides by bacterial denitrification processes – which are stimulated when a humus-depleted soil becomes water-logged.

In effect the Sanborn Field studies, and similar investigations elsewhere, were a warning that in humus-depleted soil, fertilizer nitrate tends to break out of the natural self-containment of the soil system and to penetrate the air and the water. But the warning was ignored. In 1942, when the 50-year results of the Sanborn Field studies were published, the annual U.S. consumption of chemical fertilizer amounted to somewhat less than 0.5 million tons of nitrogen. It is now about 7 million tons – a 14-fold increase in about 25 years. The vast use of nitrogen and other chemical fertilizers, together with other means of intensifying crop production, has, of course, given us a remarkable increase in yields. But as shown by the Sanborn tests, we must look for some of the 7 million tons of nitrogen added to U.S. soil during the last year not in the soil, or in the harvested crops, but in the nation's lakes and rivers, and in the rain and snow that fall on the land.

4. The Fate of Fertilizer Nitrogen

What is the present nitrogen balance in the U.S.? It should be stated at the outset that this question cannot now be answered, for we simply lack the data needed to draw up a total balance sheet. Overall statistics are misleading. The total amount of nitrogen temoved from U.S. soils in crops annually is now about 9–10 million tons [2]. Since this figure is about 2–3 million tons greater than the total amount of nitrogen added to the soil in fertilizer, it is evident that some nitrogen is being withdrawn from the organic store in the soil. But it would be a mistake to conclude that *all* of the fertilizer nitrogen appears in the crops and that only 2–3 million tons of nitrogen is withdrawn from the soil's organic store. Numerous laboratory and field experiments show that only part of fertilizer nitrogen is taken up by crops. Efficiency of nitrogen uptake varies with the crop, growing conditions and nature of the soil; in no case is efficiency greater than about 80%, and values of the order of 50% are common [3]. It must be assumed, therefore, that an appreciable portion of fertilizer nitrogen may fail to be absorbed by the crop and appears in surface waters and the air. In the absence of overall data capable of evaluating the total distribution of nitrogen, the best we can do is to examine the entry of nitrogen into surface waters and the air and to determine the extent to which such nitrogen originates in fertilizer and in other possible sources.

A particularly useful set of data is available on the nitrate content of the rivers of Illinois where for the last 25 years the Illinois State Water Survey has been making increasingly detailed studies of the quality of surface waters [4]. Illinois is intensively farmed (about 60% of the state's area is in crop land) and the system of farming involves very heavy use of inorganic nitrogen fertilizer. As in most parts of the U.S., the use of inorganic nitrogen fertilizer has increased by an order of magnitude in the last 20–25 years.

Figures 1 and 2 present the results of Illinois State Water Survey studies of three rivers, the Kaskaskia, the Sangamon, and the Skillet Fork. The two former streams drain a typical intensively-farmed, heavily fertilized area; the Skillet Fork drains an area which is not heavily farmed and in which fertilizer use is minimal. Figure 1 reports the mean nitrate concentrations found in these rivers for several periods since 1945. The Kaskaskia and Sangamon rivers now show a striking monthly variation in nitrate concentration; nitrate levels reach a maximum in May and June and a minimum in September and October. This effect is closely associated with the amount

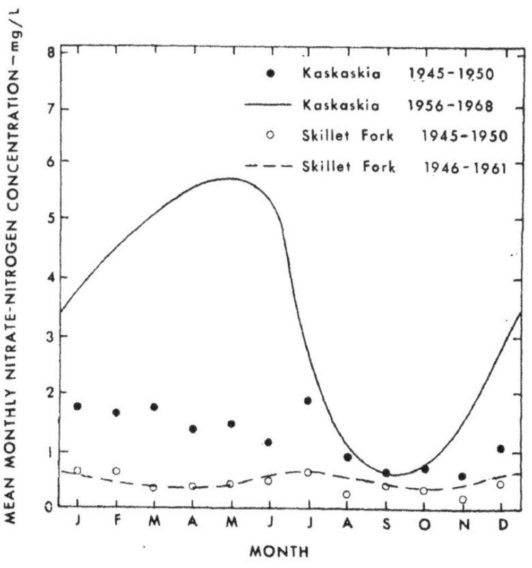

Fig. 1.

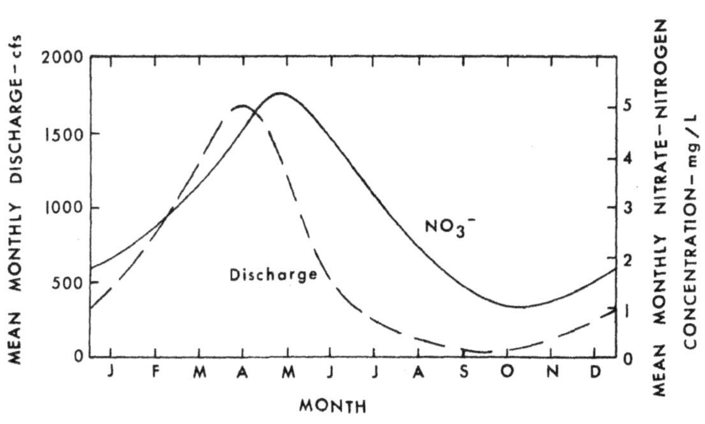

Fig. 2.

of water discharged into the rivers, suggesting that that movement of water precipitated on the land into the river is responsible for the appearance of nitrate in the stream. There is a two-month delay between the period of peak water discharge (March–April) and peak nitrate concentration (May–June). This suggests that the nitrate reaches the river by way of percolation into the soil and as a result of drainage, rather than in the immediate runoff from precipitation. This is in keeping with expectations based on the known effects of precipitation on the movement of nitrate in soil; as rain begins to penetrate the soil it carries free nitrate down into the soil toward the ground water, so that, when the soil becomes water-logged and rain runs off at its surface, little nitrate is carried off to the rivers.

It is evident from Figure 1 that the nitrate level of the Kaskaskia River has increased about threefold between 1946–50 and 1956–68. In contrast, there has been no significant change in the nitrate load of the Skillet Fork River in that period. The only known difference in the nitrogen inputs between these two drainage areas is the sharp increase, during this period, in the use of nitrogen fertilizer in the Kaskaskia area as compared with the Skillet Fork area. Hence it is likely that the increased nitrogen load of the Kaskaskia is due to the increased use of nitrogen fertilizer in its drainage area.

Other possible sources of stream nitrogen are animal wastes from feedlots, urban sewage, and nitrogen compounds present in rain and snow. Feedlots are relatively scarce in Illinois and may be disregarded as a significant source of stream nitrogen. Precipitation can also be ruled out as an important source of river nitrate, since the average nitrate-nitrogen content of precipitation in that area is about 0.1–0.2 ppm [5], whereas the Kaskaskia and the Sangamon exhibit nitrate-nitrogen concentrations in excess of 5 ppm.

The recent interim report from the Illinois State Water Survey [4] includes data which enables one to estimate the contribution of sewage nitrogen to the nitrate content of the Sangamon River. Since the population involved in sewage effluent delivered to the Sangamon River is known (7000–8000 persons), as is the average nitrogen contribution to sewage per capita per day (about 0.11 lbs. of nitrate), it is possible to calculate the expected daily contribution of sewage nitrogen to the total nitrate load of the Sangamon River: 175–200 pounds of nitrate-nitrogen per day. Figure 3 reports the calculated nitrate content carried by the Sangamon River at Monticello in pounds of nitrate-nitrogen/day. It is evident that sewage can account for the Sangamon River nitrate load only in October. The total annual nitrate-nitrogen content of the Sangamon River water is about 50 million pounds. Hence, the

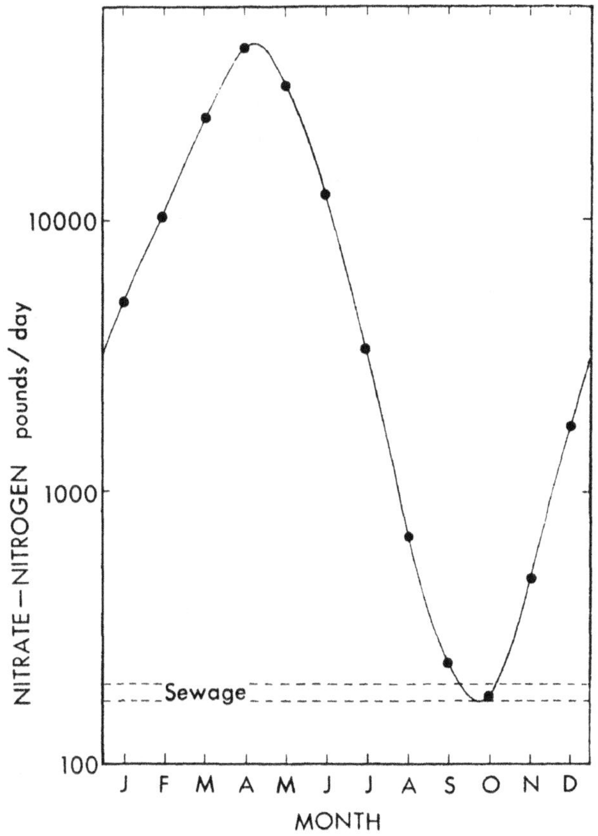

Fig. 3.

total annual nitrate-nitrogen contributed by sewage (ca. 65000 pounds) represents an insignificant proportion of the total nitrate-nitrogen entering the river. Similar calculations for the Kaskaskia River show that in 1968 nitrogen due to sewage effluent entering the river can account for less than 1% of the total nitrate-nitrogen carried by the river.

What are the consequences of the nitrate levels which have been induced in Illinois surface waters by intensive nitrogen fertilization? The Illinois Water Survey data show that all of the State's rivers which traverse heavily farmed land – and this comprises all but a few of the State's streams – have nitrate levels which are far in excess of those which lead to heavy algal growth (this level is usually estimated at about 0.3 ppm of nitrogen) [6]. Indeed, the survey has recently announced that the State's streams, as a whole, have now become eutrophic – so burdened with nutrients as to support algal blooms, which impose a sufficient organic load on the water to deplete its oxygen supply. This situation is dramatically illustrated by estimates of the nitrate content expected in a new reservoir which is about to be filled at Shelbyville on the Kaskaskia River. The interim report points out that had the dam been closed early in 1968, "it could have been filled very rapidly during April, May, and June of 1968 – with water in which the nitrate content ranged between 19.5 and 56 mg/l (i.e., 4.5–13.0 ppm of nitrate-nitrogen)." When this new reservoir is filled, there will be an immediate problem from nitrate-induced eutrophication. From the evidence cited above, it is apparent that eutrophic pollution of surface waters in Illinois, which is now critical, involves nitrogen which is largely due to fertilizer.

Illinois also faces a health hazard due to the high nitrate levels of most of its rivers. The acceptable limit for the nitrate-nitrogen concentrations of potable water established by public health agencies is 10 ppm. As noted in the statement quoted above, this level has already been exceeded by the Kaskaskia River. According to information provided to me by the Department of Health of Decatur, Illinois, the city water supply, which is taken from the Sangamon River, has approached this level during the spring months of the last two years. It has also been reported that about 25% of Illinois groundwaters from shallow wells of 25 ft. depth or less, contain more than 10 ppm of nitrate-nitrogen and that groundwater derived from fertilized fields frequently contains over 14 ppm of nitrate-nitrogen in the spring months [7]. It would appear, therefore, that in rural regions of Illinois the intensive use of nitrogen fertilizer has created a public health problem due to high levels of nitrate in potable water (the medical aspects of this problem are to be discussed below).

The U.S. Geological Survey reports water quality data, including nitrate concentrations, from a large number of river sampling stations throughout the U.S. These data represent a vast resource of information applicable to our problem, but unfortunately little has been done, thus far, to analyze them in terms which are relevant to the nitrogen cycle. As an initial step in this direction I have made a preliminary analysis of the USGS data on part of the Missouri River drainage basin [8].

Fig. 4.

At its headwaters, the Missouri River drains forested mountain regions which merge into pasture land in the Dakotas. Further downstream, the river drainage basin includes intensively farmed and heavily fertilized crop lands in Nebraska. The contributions of the two types of land areas to the river drainage can be separated by comparing data taken at Nebraska City, Nebraska, with data from upstream points at the boundary between the two areas, Williston, N.D. and Overton, Nebraska (see map, Figure 4). By this means, one can calculate both the water discharge and the nitrate specifically arising from the Nebraska agricultural drainage basin.

Calculations have been made of the mean annual nitrate concentration (weighted according to water discharge rate) and the mean annual water discharge for the Nebraska region for the period 1950–64. In general, nitrate concentration is correlated with total water discharge, a relationship which is also evident in the Illinois data. Thus, in the period 1950–55 there was a sharp decline in water discharge and concurrently a sharp decline in mean nitrate concentration. In the period 1955–64 both discharge rate and nitrate concentration increased.

From such data it is possible to calculate the total nitrate input into the river from the Nebraska drainage area. This is shown in Figure 5, both for water leaving the river at Williston, and for the Nebraska drainage area. One striking result is that the nitrogen drainage from the *upstream* parts of the Missouri River has been declining steadily since 1950. I am informed that pasture land in this region has been considerably improved by conservation practices in this period; the decline in nitrogen drainage from this area very likely reflects this effect. In any case, this curve shows that in an area which does not receive an appreciable man-made nitrogen input (it is sparsely populated and uses relatively little fertilizer), nitrogen losses from the soil are minimal,

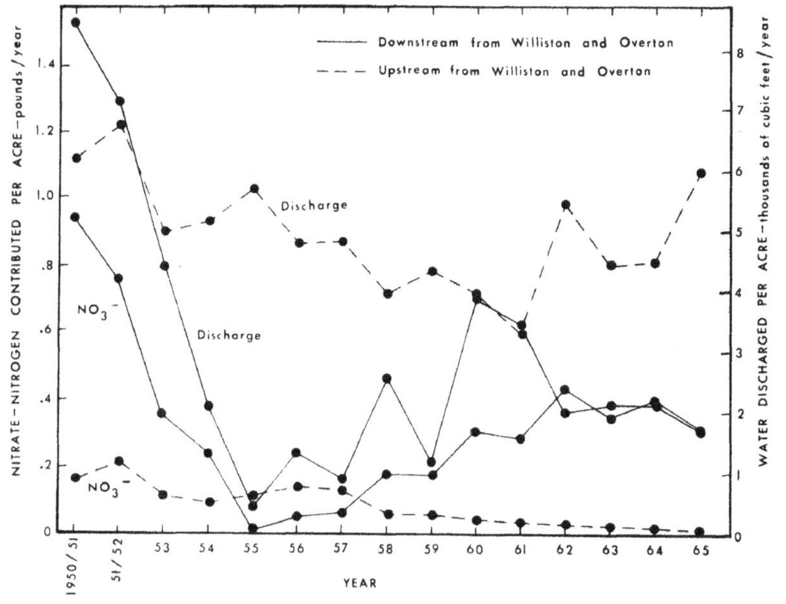

Fig. 5.

and are even capable of being reduced. In contrast the nitrate entering the Missouri River from the Nebraska area has increased sharply since 1955, following the decline due to falling water discharge. Thus, somewhere in this area there must be significant sources of inorganic nitrogen which have increased since 1955.

Figure 6 compares the annual fertilizer nitrogen tonnage used in the area with the nitrate load in the river (corrected for variations in water discharge). It is evident that

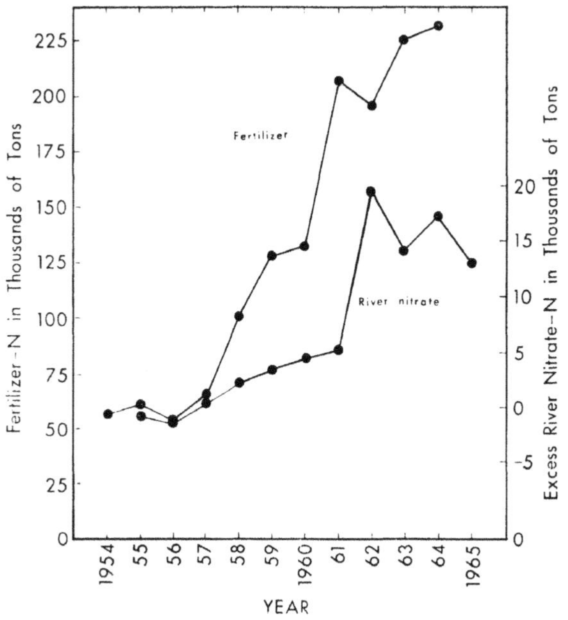

Fig. 6.

in general the nitrate load in excess of that which can be accounted for by the concurrent rate of water discharge increases during the period of rising fertilizer usage. Examination of this figure also suggests that there is a one-year delay between the effect of a change in fertilizer use and the change in nitrate load of the river. This effect is examined explicitly in Figure 7 which is a plot of the annual *change* in nitrogen fertilizer use against the annual *change* in river nitrate load *during the following year*. There appears to be a positive correlation between the two values, with the regression line passing through the origin. The regression coefficient for these data indicates, at the 99% level of confidence, that the change in nitrate load acquired by the Missouri River in the Nebraska region is proportional to the change in the amount of nitrogen-fertilizer used in the preceding year. This is the relationship expected if fertilizer nitrogen percolates into groundwater and reaches the river by that route, for this process involves a delay of some months.

Apart from fertilizer, other possible inputs of nitrogen to river water are precipitation, sewage, and livestock manure. Nitrogen concentrations in precipitation are about an order of magnitude too low to account for river nitrate levels. The relative importance of the contribution of nitrogen from sewage and manure to the river can be evaluated by examining the ammonia content of the river, for it is well-known that this substance reflects the input of organic waste. Unfortunately, ammonia analyses are usually not carried out by the USGS. Such values are obtained by public health

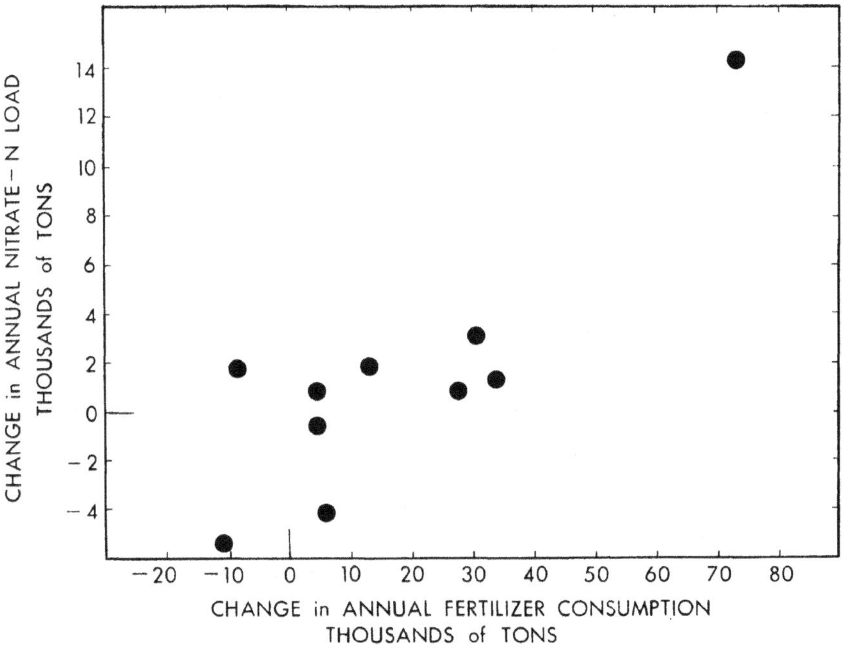

Fig. 7.

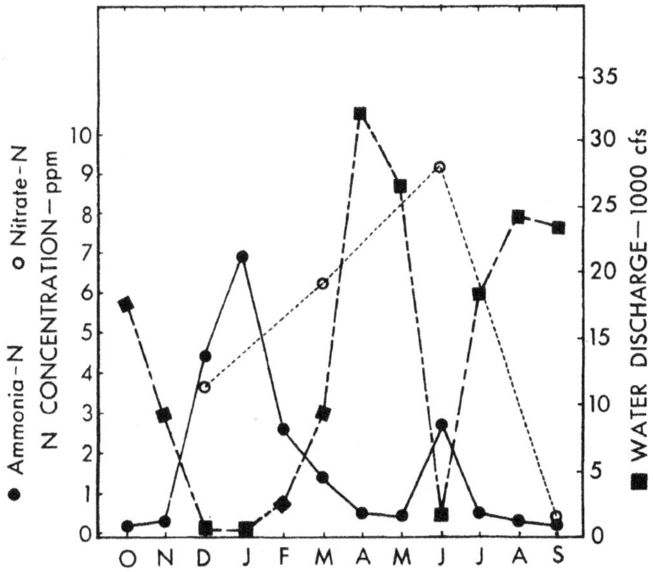

Fig. 8.

agencies (which, for their part, usually fail to analyse river nitrate levels). However, the U.S. Public Health Service has reported [9] ammonia values at river stations quite close to the USGS stations at Williston, Overton, and Nebraska City, so that the concentrations of ammonia and nitrate in the water added to the river in the Nebraska drainage area can be compared.

Seasonal changes in these values, averaged for the period 1957–63 are shown in Figure 8. The ammonia concentration is *inversely* related to the seasonal rate of water flow. This is the relationship expected if the ammonia were derived from material (such as sewage and runoff from manured lands) added directly to the river at a rate *independent* of water flow, and therefore diluted by the latter. In contrast, nitrate concentration is highest during the period of peak water flow – a relationship consistent with the conclusion that nitrate is carried into the river by the groundwater contribution to water flow. This is evidence that an appreciable part of the ammonia and nitrate entering the river in the Nebraska drainage area originates from different sources. Since ammonia is largely due to sewage effluent, livestock manure, and other organic sources, it would appear from these considerations, that a significant part of the river nitrate must originate from some other source. The only known source which remains to account for the observed nitrate content added to the Missouri River in the Nebraska region is fertilizer.

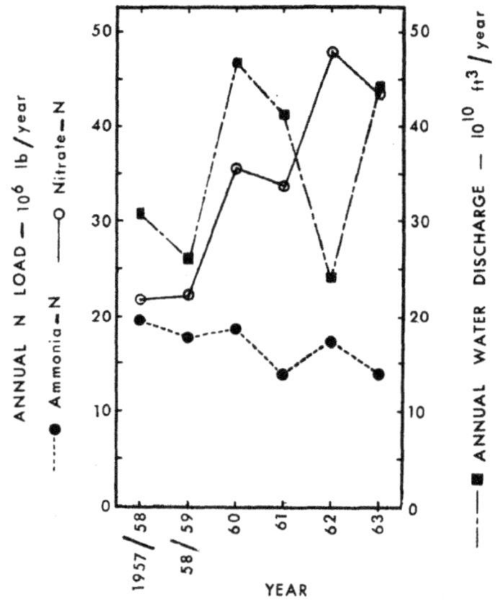

Fig. 9.

Additional evidence in support of this conclusion is shown in Figure 9, which compares the ammonia and nitrate concentrations of the water added to the river in the Nebraska region, annually, for the period of 1957 to 1963. Ammonia concentrations do not change significantly in that period; in contrast, nitrate concentrations increase appreciably. Thus, while changes in annual nitrate concentration tend to parallel concurrent changes in fertilizer consumption, changes in ammonia concentration do not. Again, this indicates that river nitrate levels are related to input from fertilizer rather than from sewage and manure.

Thus, as in the case of the rivers of Illinois, this analysis leads to the conclusion that inorganic nitrogen fertilizer contributes significantly to the nitrogen content acquired by the Missouri River from drainage arising in the Nebraska region.

An important consequence of these analyses is that they provide specific evidence that nitrate originating in fertilizer reaches surface waters by a process which carries it from the soil in proportion to the rate of flow of surface waters. In contrast, nitrogen originating in sewage effluents is delivered to the stream at a rate independent of water flow.

Against this background, it is useful to take note of the results of a detailed survey of the nitrogen sources which contribute to the nutrient levels of Wisconsin surface waters – many of which are now eutrophic – reported by the Water Subcommittee of the Wisconsin Natural Resources Committee of State Agencies in 1967 [10]. According to this report, 42% of the nitrogen reaching Wisconsin surface waters originates in groundwater; 30% originates in urban and private sewage systems; about 10% is due to runoff from manured crop land; 9% is due to nitrogen in precipitation falling on water areas. This suggests strongly that fertilizer is a major source of the nitrate which contributes to the eutrophication of Wisconsin surface waters. Similar calculations can be derived from the recent Lake Erie Report of the Federal Water Pollution Commission [11]. The report estimates that the lake receives about 75 million pounds of nitrogen per year from the surrounding farmlands. Municipal sewage contributes about 90 million pounds of nitrogen per year. Thus urban and agricultural sources contribute about equally to the nitrogen pollution of Lake Erie. A separate calculation, based on the expected losses of fertilizer nitrogen leaching from the crop land of the Lake Erie drainage basin [12] yields a similar result. These observations suggest that about one-half of the nitrate which pollutes such surface waters (which are in areas that include major urban sources of sewage) is derived from fertilizer nitrogen.

It has been claimed, in the past, by some investigators [13] that high nitrate levels in groundwater wells in rural areas is nearly always due to contamination from sewage systems or livestock feedlots. Recently this problem has been investigated in considerable detail in an intensely farmed irrigated area of the South Platte River valley in Colorado [14]. This study confirmed that groundwater under feedlots was usually contaminated with very high concentrations of nitrate. However, high nitrate levels (about 10–30 ppm of nitrate-N) were also found in groundwater beneath irrigated fields, and it was calculated that 25–30 pounds of nitrogen per acre drained from these fields annually. The report concludes:

> The ratio of irrigated land to that in feedlots for the study area was approximately 200 to 1. Therefore, although much larger amounts of nitrate per unit area were usually present under feedlots, indications were that irrigated lands contributed more total nitrate to groundwater. Feedlots, however, are usually located near the homestead and may have a pronounced effect on the water quality of domestic wells.

The Colorado study provides an important way to distinguish between pollution due to feedlots and fertilizer: in the former, nitrate is always accompanied by comparable – or greater – concentrations of ammonia and by organic materials. Unfortunately, such data are not available from most of the older studies which claim that groundwater contamination is due to feedlots and not to fertilizer [15]. Data on the nitrate content of groundwater in other parts of the country support the general conclusion that fertilizer is a major souıce of this contaminant. Thus, McHarg's studies of groundwaters in Minnesota [16] show that nitrate-nitrogen levels of the order of 10 ppm are common, that nitrate originates at the surface of the soil, and is especially concentrated in groundwater under intensely farmed and fertilized areas. In at least one Minnesota town, Elgin, the city's water supply has exceeded the 10 ppm public health limit, and has had to be replaced [17].

The heavily farmed areas of California probably represent the most severe public health hazard from nitrate originating in fertilizer. These agricultural areas are heavily farmed, usually under irrigation, with a very intense and increasing utilization of nitrogen fertilizer. In a typical situation in the San Joaquin Valley it was found that fields irrigated with water containing 1.7 ppm of nitrate-nitrogen yielded, in the subsurface drainage water, nitrate levels which averaged (over a 4-year period, 1962–66) 44.5 ppm of nitrate-nitrogen. An average of 167 lbs./acre/year of fertilizer nitrogen was supplied to the soil [18]. Another recent study [19] shows that in 4 different cropping systems an average of 36% of the applied fertilizer nitrogen appears in the drainage water. The nitrate concentrations of drainage waters from different areas generally parallel the amount of fertilizer nitrogen supplied.

The irrigation drainage water in this area is at present collected by the San Joaquin River which discharges into San Francisco Bay. The nitrate content of the river water has now become so high as to induce serious water pollution, due to eutrophication, both in the river itself and in reaches of San Francisco Bay. It is evident from the above that fertilizer nitrogen contributes significantly to this environmental deterioration.

Serious health hazards have arisen in the San Joaquin Valley and in Southern California because of the high nitrate levels of wells derived from groundwater supplies. A study of 800 wells in Southern California in 1960–61 [20] showed that in 182 wells water exceeded 5 ppm of nitrate-nitrogen and in 88 wells nitrate-nitrogen exceeded 10 ppm (the public health limit for potable water). This study provides explicit evidence that the nitrate contamination was not due to organic sources such as sewage or manure. While nitrate-nitrogen levels were in the range 1–10 ppm, the concurrent organic nitrogen levels were of the order of 0.3 ppm and ammonia was in the range of 0.1–0.5 ppm. From the Colorado studies described earlier it is known that groundwater contaminated by organic sources invariably exhibits ammonia and organic nitrogen levels that are in excess of the concurrent nitrate levels. Hence it must be concluded that organic sources do not contribute significantly to the nitrate contamination of wells in Southern California, so that the contributing source must be fertilizer nitrogen. In certain areas of California nitrate levels of well water is so high that public health officials have been required to warn physicians against the use of such water in infant feeding.

It appears evident from these results that in heavily fertilized areas such as California, fertilizer nitrogen is the chief source of two environmental hazards: eutrophication of surface waters and unacceptable levels of nitrate in potable water.

5. Excessive Nitrate in Foods

Under natural circumstances, before the intrusion of modern agriculture, nearly all of the soil nitrate which nourishes the plant is converted to protein and other organic plant constituents; in most plants relatively little of the observed nitrogen accumulates as nitrate. However, at the heavy rates of chemical fertilization now widely used in the U.S., this situation has been changed; plants grown on soil heavily fertilized with nitrate contain much-increased amounts of nitrate. For example, a lettuce crop grown on Missouri soil without added nitrogen contained about 0.1% of nitrate-nitrogen; given 100 lbs. per acre of nitrate fertilizer the lettuce nitrate content increased to 0.3% and at about 400 lbs. of nitrogen fertilizer per acre, the nitrate content of the crop reached a maximum of 0.6% [21]. Other possible causes of nitrate accumulation in crops are weather conditions, light intensity, and time of harvest.

In contrast to the well-reported instances of methemoglobinemia arising from nitrate-polluted drinking water, there are, to my knowledge, in *the United States* no clinical reports of methemoglobinemia relating to nitrate levels of foods. However, in the last five years, European pediatricians have reported a sufficient number of cases of methemoglobinemia associated with ingestion of spinach to warrant a recent general review [22]. In these cases the spinach preparations fed to the affected infants were found to contain sufficient *nitrite* to cause the poisoning. However, *fresh* spinach, even though often high in *nitrate*, contains only insignificant amounts of nitrite, so that the toxic nitrite levels must have developed in the food somewhere between harvesting and use. This process has now been explained by the work of Schuphan [23]. When nitrate-rich spinach preparations are exposed to certain common air-borne bacteria, the latter can quickly convert the nitrate to toxic amounts of nitrite. This may happen when a jar of spinach baby food is opened to the air and kept for a day or two (even in the refrigerator) before using. Bacterial nitrite production may also occur when frozen spinach is kept in the thawed condition for a day or two, either during storage of the original package, or after the package has been opened. Even in the absence of bacterial action, when spinach is transported in tightly packed containers, enzymes in the spinach itself may convert nitrate to nitrite.

On the basis of these generalizations Simon [22] has made the following recommendations regarding the use of spinach for infant feeding:

(1) Prepared spinach must not be kept at room temperature.

(2) Too much fertilizer should not be used for growing spinach. (This is discussed further below.)

(3) Spinach used for infant feeding must not contain more than 300 mg nitrate per kg* because

* mg/kg is equivalent to ppm.

of the danger of possible mistakes in storage, for nitrite in toxic amounts can only be formed if there is a high nitrate content.

(4) Spinach should not be given to infants in the first 3 months of life, for at this age nitrate can be reduced in the upper parts of the gastrointestinal tract, and because of the lowered diaphorase activity there is then increased susceptibility to methemoglobinemia.

These guidelines provide a useful background for evaluating the significance of observed nitrate levels of foods.

The most detailed study of nitrate in baby foods in North America appears to be that of Kamm *et al.* [24]. They report average values based on the analysis of a total of 37 samples of commercial baby foods in Canada (see Table I).

TABLE I

	Average ppm of nitrate	No. of samples analyzed
Mixed vegetables	88	2
Carrots	101	8
Green beans	163	3
Garden vegetables	180	5
Graham crackers	211	1
Squash	282	5
Wax beans	444	2
Beets	977	6
Spinach	1373	5

Baby foods not containing vegetables gave considerably lower nitrate values. A less complete study of baby foods reported earlier by the Missouri Agricultural Experiment Station [21] gave results similar to those cited above.

The most recent baby food analyses known to me are those performed in January, 1968 by W. Brabent, Superintendent and M. Boulerice, Chemist of the Department of Health of the City of Montreal [25]. Analyzing single commercial baby food samples, they found: in spinach, 642 ppm nitrate; in beets, 523 ppm nitrate; in squash, 295 ppm nitrate.

It is clear from these results that infants fed on commercial baby foods of the types analyzed in Canada and the United States in recent years are very likely to exceed, in their diet, the nitrate level recommended by Simon as an acceptable limit.

The European pediatric literature also provides a useful insight into the conditions under which nitrate contained in baby food (specifically, spinach) may be hazardous to infants. It should be recalled that the danger is not from nitrate itself, but from *nitrite* which is readily formed from nitrate under reducing conditions. Nitrate can be converted to nitrite in several different ways:

(1) The infant's intestinal tract normally contains bacteria which can convert nitrate to nitrite. In a normal infant most of the nitrate in food is absorbed from the digestive tract before it reaches the intestine, thus minimizing the hazard. However, during digestive disturbances, bacteria from the intestine may reach the upper portions

of the digestive tract where they come in contact with nitrate and produce nitrite. It is also possible that infants may become infected with abnormal strains of intestinal bacteria which are particularly active in converting nitrate to nitrite. (A 1962 epidemic of infant methemoglobinemia in a French hospital was traced to such an effect [26].)

(2) Fresh spinach may, during transport and storage, convert nitrate to nitrite by enzymatic action. Thus Phillips [27] reports that analyses of fresh spinach with an initial nitrate content of 2314 ppm showed a complete conversion of nitrate to nitrite after 21 days of refrigerator storage; in 14 days two-fifths of the nitrate was converted to nitrite. This effect is not uniform and is related to the original nitrate content of the spinach. Similarly, Schuphan [23] reports that toxic levels of nitrite are produced when nitrate-rich spinach is tightly packed in crates and transported and stored over a 4-day period.

(3) When frozen spinach is thawed, bacteria originally present in the spinach may convert nitrate to nitrite. Thus Phillips [27] reports that 3 samples of frozen spinach allowed to thaw over a 39-hour period contained 99, 63, and 244 ppm of *nitrite* respectively. Sinios and Wodsak [28] report an instance in which frozen spinach, thawed and stored at 37 °C for 48 hours contained an average of about 400 ppm of nitrite. It should be noted that frozen spinach might readily be subjected to a thawing period of this duration during storage accidents.

(4) When a sterile jar of baby food spinach puree is opened to the air, bacteria capable of converting nitrate to nitrite may enter the food. Phillips [27] tested several jars of baby food in this way and reported no significant increase in nitrite. However, it is obvious that this effect depends considerably on the degree of local sanitation (i.e., prevalence of dust and contamination from intestinal bacteria) so that one cannot rely on the safety of a jar of nitrate-rich baby food once it has been opened. In a number of the European clinical cases of infant methemoglobinemia, it is evident that nitrite was produced in stored spinach puree by bacterial action (see Simon [22]). This accounts for the warning by European pediatricians against the use of stored containers of baby food spinach.

From the foregoing it is evident that at least several common baby food preparations, spinach, squash, beets, and wax beans, may contain nitrate levels of 300 ppm or more, which, on the basis of European clinical reports, can lead to methemoglobinemia. While such a high level of nitrate in the original food is a *necessary* condition for the occurrence of methemoglobinemia, it is not a *sufficient* condition, for the nitrate must be converted to nitrite, by one of the means described above, before reaction with hemoglobin occurs. Thus, eating nitrate-rich food creates a definite risk of methemoglobinemia. This risk can be reduced by avoiding the feeding of such foods to infants suffering from a gastrointestinal upset, by avoiding the use of frozen foods which have been thawed for a day or two, and by avoiding the use of commercial baby foods (i.e., of the suspect vegetables) which have been stored after being opened. If the original vegetable is low in nitrate, all of these risks can be avoided.

Schuphan [23] has reported observations which tie the production of toxic amounts of nitrite in spinach directly to fertilizer practice. He obtained fresh spinach harvested from land treated with different amounts of inorganic nitrogen fertilizer and studied the conversion of nitrate to nitrite during a 4-day period of transport and storage. In the case of spinach fertilized with 320 kgm of nitrogen per hectare, the leaves contained about 3500 mgm of nitrate per 100 gm dry weight and no nitrite at harvest; after the 4-day period the spinach contained about 360 mgm of nitrite per 100 gm dry weight and the nitrate level had fallen about 30%. Spinach fertilized with 80 kgm of nitrogen per hectare contained at harvest about 500 mgm of nitrate per 100 gm and no nitrite; after transport and storage no changes in either nitrate or nitrite content were observed. On the basis of these data Schuphan concludes that the medical hazard of spinach in infant feeding arises from the excessive use of nitrogen fertilizer. Simon [22] ascribes the recent appearance of infant methemoglobinemia to the shift, especially in Germany, from spinach production by small market gardeners to large-scale 'industrial' farms, which, according to him, are "totally lacking in experience with fertilization".

Unfortunately, there appear to be no detailed studies of the recent trends in the actual levels of nitrogen fertilizer used in the production of specific vegetable crops in the United States, nor comparable year-to-year studies of the resultant nitrate levels in the crops. Jackson *et al.* [29] have made a single comparison of nitrate contents of various vegetable crops grown in 1964 with data for 1907 crops reported by Richardson [30] and find no significant differences. However given the differences in the geographical origins of the 1907 and 1964 crops, lack of information as to the fertilizer practices actually used in growing the crops, and differences in methodology, no firm conclusions can be based on this single comparison.

In contrast, indirect evidence does indicate that the nitrate contents of vegetables grown in the United States have been increasing and that this can be traced to more intensive use of nitrogen fertilizer. Thus Johnson [31] reports that the nitrate content of tomatoes is directly related to the amount of nitrogen fertilizer used. At 328 pounds of nitrogen per acre, tomatoes contained 39 ppm of nitrate; with 408 pounds of nitrogen per acre, nitrate rose to 57 ppm; at 488 pounds of nitrogen per acre, nitrate was 97 ppm; at 568 pounds of nitrogen per acre, nitrate was 144 ppm. Johnson found that tomatoes containing 53–78 ppm of nitrate were corrosive to the tinplate of canning containers. Since nitrate-induced corrosion of tinplate by canned vegetables is a phenomenon which has been of concern to canners only in the last few years, we can infer that recent increases in the use of nitrogen fertilizer are indeed causing comparable increases in the nitrate content of commercial vegetable crops. The increasing importance of nitrate poisoning in livestock, which has been widely noted, also suggests that the nitrate levels of crops which are used in livestock feeds have been increasing in recent years.

6. Nitrogen Compounds in the Air

That the atmosphere now contains significant amounts of ammonia and nitrogen oxides is revealed by extensive analyses of the nitrogen content of rain and snow [32].

These compounds are washed out of the atmosphere by precipitation and in the process nitrate (and very little nitrite) is formed both from nitrogen oxides and probably by oxidation of ammonia. Formation of ammonia from nitrogen gas in the atmosphere appears to be highly unlikely. The observed concentration of nitrogen oxides in the air cannot be accounted for by lightning discharges [33]; however, it is possible that some photochemical fixation of nitrogen gas does occur. Generally, natural atmospheric processes cannot account for the observed concentrations of ammonia and nitrate in precipitation. Most of these materials clearly enter the atmosphere from the land. At the coastline and over the ocean their concentrations in precipitation are very low, but over land masses appreciable concentrations of ammonia and nitrate are found in collected precipitation.

Regional and seasonal variations in the inorganic nitrogen content of precipitation are quite striking and provide useful clues as to the origins of these materials. Thus, at Frankfurt-on-Main, Germany, two seasonal maxima in atmospheric nitrogen dioxide are observed. One of these, in April, appears to be due to biological processes in the soil; the second maximum, in December and January, is associated with fuel combustion. In contrast, at a rural station in Germany, only a single maximum, in April, due to soil activity, was observed [34].

In order to evaluate the significance of technological intrusions in this segment of the nitrogen cycle, we need to consider the possible effects of two activities on the concentration of nitrogen compounds in precipitation – nitrogen fixation due to automobiles and certain industrial activities, and fertilizer. As indicated earlier, it is well-known that automobile exhausts and industrial combustion processes emit significant amounts of nitrogen oxides and that part of the nitrogen present in the soil may be released into the air as ammonia and nitrogen oxides. Regional variations in

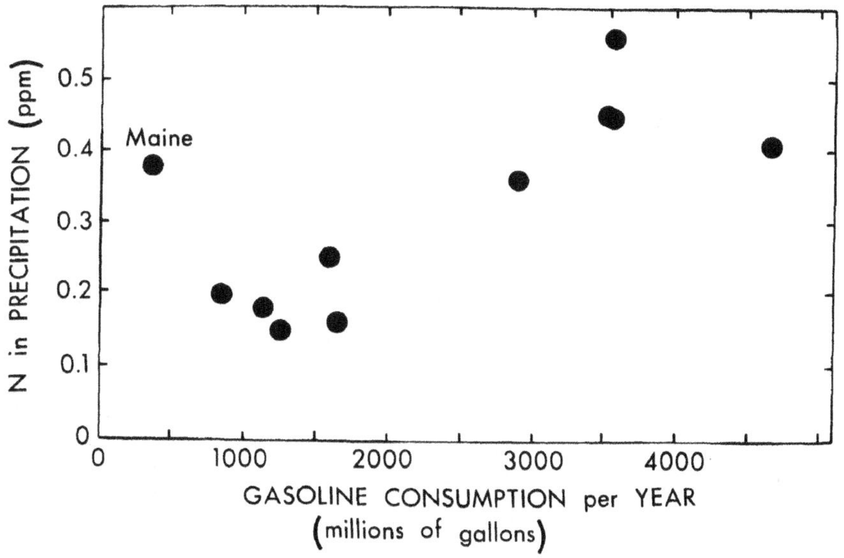

Fig. 10.

the nitrate content of precipitation over the U.S. suggested to Junge [35] the influence of both of these technological effects. He observed high concentration of nitrate over heavily populated and industrialized regions which appear to reflect nitrogen fixation in automobile engines and combustion processes. He also notes high seasonal concentration of ammonia and nitrate over agricultural regions (California, Texas, the Corn Belt) which appear to coincide with the time of intensive use of fertilizer.

Recently Lodge *et al.* [36] have reported on the inorganic nitrogen concentration in precipitation from various regions of the U.S. for the period 1960–66. When these results are compared with the regional use of gasoline (as an index of automotive activity) in areas which use relatively little nitrogen fertilizer, the results shown in Figure 10 are obtained. It is evident that the observed inorganic nitrogen content of precipitation correlates well with the regional use of gasoline.

7. The Nitrogen Cycle Outside the U.S.

This discussion has necessarily concentrated on the situation in the U.S., where the needed data are, though very inadequate, relatively available. That the fertilizer problem exists in European countries is evident from the German experience with methemoglobinemia from food-borne nitrate, and from well-known evidence of rapidly increasing eutrophication of surface waters. Japanese reports, where fertilizer use is very intensive, also are indicative of a severe eutrophication problem in recent years. Special problems must be anticipated in tropical and arid regions. That fertilizer practice results in a serious intrusion on the nitrogen cycle in arid regions is evident from Yaalon's recent report [37] on the inorganic nitrogen content of precipitation in Israel. His data for 1962–63 were about one order of magnitude above observations made in 1922–24. This increase in the inorganic nitrogen content of precipitation is attributed to the sharp increase in the use of fertilizer in Israel during the last 40 years. In fact, Yaalon reports that the total quantity of inorganic nitrogen now deposited annually by precipitation in Israel is about equal to the amount of nitrogen deposited on the land as fertilizer. In effect, the total mass of inorganic nitrogen artificially introduced into the cycle is now carried by the atmosphere.

8. Some Conclusions

This paper has been an inquiry into the state of the nitrogen cycle, chiefly in the U.S., as it has been affected by recent technological intrusions. I believe that the evidence supports the following conclusions:

(1) The natural nitrogen cycle in the U.S. has been stressed by the growth of urban populations in the last generation. However, the stress on the cycle due to the technological introductions of inorganic forms of nitrogen is considerably more severe. This is true because (a) the technological stress has grown much more rapidly than the biological one, and is now about equal in magnitude to the nutritional cycle as a whole; (b) the technological stress intrudes on the natural cycle at its most vulnerable

point by introducing into it oxidized forms of nitrogen, which can lead to environmental hazards.

(2) There are two main technological stresses on the nitrogen cycle: the introduction of inorganic nitrogen fertilizers into the soil, and the introduction into the air of nitrogen oxides due to automobiles and other combustion processes.

(3) The massive addition of inorganic nitrogen fertilizer to the soil follows a long period in which the organic nitrogen of the soil has been reduced so that nutrient uptake has become relatively inefficient. With the continued increase of fertilizer application, inorganic nitrogen – chiefly nitrate – has been leached from the soil, entering groundwater, and eventually, surface waters and wells.

(4) In heavily farmed areas, nitrate due to fertilizer makes a major contribution to the inorganic nitrogen content of surface waters, and in many places is responsible for water pollution through eutrophic stimulation of algal growth. This process is now a serious source of environmental pollution.

(5) In heavily farmed areas, the nitrate level of surface waters and wells often exceeds the public health standards for acceptable potable water, resulting in a risk to human health from methemoglobinemia.

(6) Some vegetable products in the United States and Canada, including certain baby foods, with a significant frequency, exceed the nitrate levels recommended for infant feeding by pediatricians. European studies indicate that excess nitrate levels in such products are usually the result of intensive use of nitrogen fertilizer.

(7) The inorganic nitrogen content of precipitation has increased noticeably in the last 25–30 years due to nitrogen oxides released from the soil and produced by automobiles and other combustion processes.

In my view these conclusions confront us with a number of problems which are large in their magnitude, difficult in their complexity, and grave in their import for the nation. In sum, we have in the United States, thrown the nitrogen cycle seriously out of balance. As a result, the oxidized forms of nitrogen, which are in nature maintained at low, steady-state concentrations in air, water, plants and animals have been elevated to levels which threaten the integrity of major processes in the ecosystem, and vital biological processes in animals and man. The present stress on the nitrogen cycle has already produced important environmental hazards and carries the risk of equally serious medical hazards. Clearly, corrective measures are urgent.

If correction of the present imbalance in the nitrogen cycle required only the development of the necessary technical steps, the problem would be difficult enough. Unfortunately, these difficulties are enormously magnified by the fact that the practices which have produced the imbalance are now deeply embedded in the nation's economy. This is evident from their origin in major economic activities: agriculture, industry and transport, and urbanization. The seriousness of the issues which arise from the impact of these activities on the nitrogen balance is perhaps best illustrated by the problem of water pollution from eutrophication. As I have already indicated, urban wastes and farmland drainage are probably about equally responsible for the eutrophication problem in many areas of the nation. Technological means are available for controlling the input of inorganic salts from municipal and industrial waste; essentially this involves the installation of some type of desalting process as a tertiary stage in treatment plants. However, farmland drainage is not so conveniently localized, but reaches surface waters from every shoreline point. There is no conceivable means of recovering such drainage materials before they reach the water. Similarly, there is little possibility of removing excess nitrate from food. I believe that these difficulties will eventually require the *limitation* of the current high rate of use of inorganic nitrogen fertilizer. I hardly need emphasize the explosive consequences of imposing such a restriction on the current farm economy. Intensive use of fertilizer has become a major source of farm income, for in the last 25 years the costs of land, labor, and machinery have all risen considerably – but the cost of fertilizer has dropped. The farmer's income has become crucially dependent on the intensive use of fertilizer, particularly nitrogen. If, as I believe, it becomes necessary to limit the use of inorganic nitrogen fertilizer, the present system of farming is faced with a massive dislocation.

Various palliative measures are possible: the use of slow-release fertilizers, more careful seasonal timing of fertilizer applications, precise determination of the amount and rate of fertilizer application. However, such measures do not get at the root of the problem – the continuing depletion of the organic nitrogen content of the soil, and with it, of the soil's efficiency in transferring nutrient to the crop. Attention to this fundamental problem is essential, I believe, if we are to attain a long-term equilibrium in the nitrogen cycle. This will require a serious revision in our approach to agricultural production – for example, the development of new ways to return organic matter to the soil (which would, of course, at the same time alleviate water pollution due to urban wastes), and more intensive use of soil-building pasturage systems. In turn, such an approach will confront us with major economic difficulties for it will require a relatively slow yield from agricultural investment, in contrast with present efforts to find more rapid ways to regain the farmer's capital investment. This approach will also confront the economic inertia due to the present investment of the chemical industry in fertilizer production. At the present time the nitrogen fertilizer industry is in a state of over-capacity and is seeking new ways to sell its product – for example, large scale use of nitrogen to fertilize timber crops. If this is done our present environmental problems will become worse.

What can we do to resolve these issues? Clearly much more needs to be done. Most important is the lack of detailed, continuing surveys of the nitrate and nitrite content of surface waters and their relation to contributions from fertilizer and organic wastes. Although the frequency of such surveys has begun to increase recently, they are usually not sufficiently detailed to yield an assessment of the added nitrogen among the various possible sources. Also important are continuing national surveys of nitrogen in rain and snow – in effect a monitoring system which would parallel the very effective national survey of radioactivity operated by the U.S. Public Health Service. Particularly urgent, it seems to me, is the need for detailed national surveys of nitrate and nitrite in food and drinking water, especially as they affect infants. With such information in hand it should be possible to determine the extent of the hazard from excess nitrate, both to human health, and to the stability of the self-purifying systems in surface waters, and to estimate the degree to which such hazards arise from agriculture, from urban and industrial wastes, and from the emissions of combustion plants and automotive engines.

We know from our experience with the fallout problem that even these actions will be insufficient to resolve the issue. When the fallout controversy had progressed to the point that the harmful consequences of fallout radioactivity was widely acknowledged, the conflict moved into the arena of politics. Given the acknowledged fact that fallout radiation could cause some number of genetic defects and cancers in the populations, it could nevertheless be argued that this risk was justified by the over-riding importance of developing nuclear weapons to secure the nation's military strength. There is, of course, no scientific means for resolving such a conflict. A choice which balances some number of leukemia cases against the development of a new nuclear weapon must reflect the value which we place on human life, and our belief in the wisdom and morality of relying on nuclear weapons for the security of the nation. These decisions are, necessarily, moral and political judgments. They comprised the substance of the great public debate that led to the adoption of the Nuclear Test Ban Treaty.

We can expect the same kind of public debate in connection with the artificial intrusion of nitrogen into the biosphere. Suggestions, such as those which I have made here, that we urgently need more information on the problem, are likely to confront the inertia generated by past acceptance of our present agricultural practices. If, as I expect, they will, new data finally make it clear that we need to restrict the use of inorganic nitrogen fertilizer if hazards to health and to the integrity of our surface waters are to be avoided, other objections will be raised. The world's need for food is acute and will increase. Intensive use of nitrogen fertilizer is clearly responsible for a good part of our present level of agricultural productivity. Shall we worsen world famine in order to protect the health of some portion of our infant population and the integrity of our waste-disposal systems?

There will be those who, perceiving these enormous economic, political and moral issues, will prefer to turn away from scientific discussion and public debate. The scientists who confronted the fallout problem, heard such counsel – and rejected it. They chose to speak out, in the profound conviction that the decisions were not theirs to make, but belonged to the people as a whole.

There is a unique relationship between the scientist's social responsibilities and the general duties of citizenship. If the scientist, directly or by inferences from his actions, lays claim to a special responsibility for the resolution of the policy issues which relate to technology, he may, in effect, prevent others from performing their own political duties. If the scientist fails in his duty to inform citizens, they are precluded from the gravest acts of citizenship and lose their right of conscience.

Every major advance in the technological competence of man has enforced revolutionary changes in the economic and political structure of society. The present age of technology is no exception to this rule of history. We already know the enormous benefits it can bestow, and we have begun to perceive its frightful threats. The political crisis generated by this knowledge is upon us.

Science can reveal the depth of this crisis, but only social action can resolve it. Science can now serve society by exposing the crisis of modern technology to the judgment of all mankind. Only this judgment can determine whether the knowledge that science has given us shall destroy humanity or advance the welfare of man.

Acknowledgment

This investigation was supported by grant P 10 ES 00139 from the Public Health Service, Department of Health, Education and Welfare through the Center for the Biology of Natural Systems, Washington University, St. Louis.

References

[1] Smith, G. E.: 1942, Sanborn Field, Bulletin 458, University of Missouri, College of Agriculture, Agricultural Experiment Station, Columbia, Missouri.
[2] Plant Food Review, Winter, 1965.
[3] Allison, F. E.: 1966, *Advances in Agron.* **18**, 219.
[4] Larson, T. E. and Larson, B. O.: Quality of Surface Waters in Illinois. Illinois State Water Survey, Urbana, Illinois, 1957; 'Interim Report on the Presence of Nitrate in Illinois Surface Waters'. Illinois State Water Survey, Urbana, Illinois, November 1968.
[5] Feth, J. H.: 1966, *Water Resources Res.* **2**, 41.
[6] 'Eutrophication – A Review', California State Water Quality Control Board, Publication #34, 1967.
[7] Larson, T. E.: 1968, personal communication.
[8] U.S. Geological Survey, Water Supply Papers Nos. 1198, 1251, 1291, 1351, 1401, 1451, 1521, 1572, 1643, 1743, 1883, 1943, 1949, and USGS Water Quality Records, Nebraska, North Dakota, South Dakota, 1964–65.
[9] National Water Quality Network, Annual Compilation of Data 1957–63, U.S. Department, H.E.W.
[10] Corey, R. B.: 1967, 'Excessive Water Fertilization', Report to the Water Subcommittee, Natural Resources Committee of State Agencies, Madison, Wisconsin.
[11] Lake Erie Report, Federal Water Pollution Control Administration, August, 1968.
[12] Commoner, B.: 1968, 'The Killing of a Great Lake', World Book Year Book, Field Enterprises Education Corp., Chicago.
[13] See for example, Smith, G. E.: 1964, 'Nitrate Problems in Plants and Water Supplies in Missouri', Contr. No. 2830, Mo. Agricultural Exp. Station.
[14] Stewart, B. A., Viets, Jr., F. G., and Hutchinson, G. L.: 1968, *J. Soil Water Conserv.* **23**, 13.
[15] See for example, Engberg, R. A.: 1967, Nebraska Water Survey Paper 21, University of Nebraska, Lincoln, Nebr., and Reference [13].
[16] McHarg, I.: 1968, Report to Twin-Cities Metropolitan Council.
[17] Rochester (Minn.) Post-Bulletin, November 21, 1967, p. 17.
[18] Doneen, L. D.: 1966, 'Effects of Soil Salinity and Nitrates on Tile Drainage in San Joaquin Valley, California', Water Sci. and Eng. Paper 4002. Sacramento, California; see also, 1968, 'San Joaquin Master Drain', Appendix, Part C., Federal Water Pollution Control Administration', Southwest Region.
[19] Johnston, W. R., Ittihadieh, F., Daum, R. M., and Pillsbury, A. F.: 1965, *Proc. Soil Sci. Soc.* p. 287.
[20] Occurrence of Nitrate in Groundwater Supplies in Southern California, Bureau of Sanitary Engineering, California State Dept. of Health, February 1963.
[21] Brown, J. R. and Smith, G. E.: 1967, 'Nitrate Accumulation in Vegetable Crops as Influenced by Soil Fertility Practices', Res. Bulletin 920, University of Missouri Agricultural Exp. Station.

[22] Simon, C.: 1966, 'L'intoxication par les nitrites après ingestion d'épinards', *Arch. Fr. Pédiat.* **23**, 231; 1966, 'Nitrate Poisoning from Spinach', *The Lancet* **I**, 872; also Simon *et al.*: 1964, *Z. Kinderheilk.* **91**, 124.
[23] Schuphan, von, W.: 1965, 'Der Nitratgehalt von Spinat (Spinacia oleracea L.) in Beziehung zur Methämoglobinämie der Säuglinge', *Z. Ernährungswiss.* **5**, 207.
[24] Kamm, L., McKeown, G. C. and Smith, D. M.: 1965, 'New Colorimetric Method for the Determination of the Nitrate and Nitrite Content of Baby Foods', *A.O.A.C.* **48**, 892.
[25] Boulerice, M.: 1968, personal communication. Dept. of Health, City of Montreal.
[26] Fandre, M., Coffin, R., Dropsy, G., and Bergel, J. P.: 1962, Epidémie de gastroentérite infantile à Escherichia coli O 127 B8 avec cyanose méthémoglobinémique', *Soc. Péd. Fr.*, Réun. Paris, **19**, 1129.
[27] Phillips, W. E. J.: 1968, 'Changes in the Nitrate and Nitrite Contents of Fresh and Processed Spinach During Storage', *Agr. Food Chem.* **16**, 88.
[28] Sinios, A. and Wodsak, W.: 1965, 'Die Spinatvergiftung des Säuglings', *Deut. Med. Wochschr.* **90**, 1956.
[29] Jackson, W. A., Steel, J. S., and Boswell, V. R.: 1967, 'Nitrates in Edible Vegetables and Vegetable Products', *Am. Soc. Hort. Sci.* **90**, 349.
[30] Richardson, W. D.: 1907, *J. Am. Chem. Soc.* **29**, 1747.
[31] Johnson, J. H.: 1966, 'Internal Can Corrosion Due to High Nitrate Content of Canned Vegetables', *Proc. Flor. State Hort. Soc.* **79**, 239.
[32] Eriksson, E.: 1952, *Tellus* **4**, 215, 271.
[33] Gambell, A. W. and Fisher, D. W.: 1964, *J. Geophys. Res.* **69**, 4203.
[34] Georgii, H.: 1963, *J. Geophys. Res.* **68**, 3963.
[35] Junge, C. E.: 1958, *Trans. Amer. Geophys. Union* **39**, 241.
[36] Lodge, J. P., Jr.: 1968, 'Chemistry of United States Precipitation', Nat. Center for Atmospheric Research, Boulder, Colorado.
[37] Yaalon, D. H.: 1964, *Tellus*, **16**, 200.

For Further Reading

1. Barry Commoner, *Science and Survival*, Viking Press, New York, 1966.
2. W. V. Bartholomew (ed.), *Soil Nitrogen*, American Society of Agronomy, No. 10, Madison, Wisc., 1966.

THE DYNAMICS OF NITROGEN TRANSFORMATIONS IN THE SOIL*

D. R. KEENEY and W. R. GARDNER

Dept. of Soil Science, University of Wisconsin, Madison, Wisc., U.S.A.

Abstract. As the nation's agriculture becomes more intensive and the need to use the soil as a waste disposal system increases, contaminations of ground water aquifers with nutrients, particularly nitrate, can be expected to become a serious and widespread problem. Already surface water supplies are showing the signs of excessive fertility while reports of high nitrate wells increase in frequency. Nitrogen compounds in surface waters often are associated with enhanced eutrophication rates while high nitrate in waters consumed by humans or animals may give rise to health problems.

1. Introduction

Concern over the pollution of ground water aquifers by nitrate has been expressed in many states. The recent literature contains references to problems in California, Missouri and Colorado while high nitrate wells were reported in Iowa, Wisconsin and Minnesota over two decades ago. The nitrogen cycle, particularly the soil phase, has been the subject of intensive study for many years. Nitrogen undergoes complex transformations between its various forms by means of chemical, physical, and biological processes. Management of this complex system and control of nitrate concentrations in ground and surface waters requires integration of research effort.

2. Environmental Problems Associated with High Nitrates in Ground Water

The most critical problem associated with nitrate levels in ground water aquifers is the possible deleterious effects on human and animal health. With animals, nitrate toxicity can result in abortions, lowered productivity, etc. With humans, only infants under six months of age not on solid foods apparently suffer lethal effects (Wright and Davison, 1964), and the problem does not appear to be serious at this time. Nitrate toxicity is caused by microbial reduction of NO_3^- to NO_2^- in the gut, and reaction of NO_2^- with ferrous iron in blood hemoglobin to ferric iron giving methemoglobin, which cannot transport oxygen.

High nitrates in irrigation water can be detrimental to agriculture. Quality and yield of crops such as grapes and sugar beets can be reduced if excess amounts of N are applied at the wrong time (Stout and Burau, 1967). Industrial processes also may be adversely influenced by high N concentrations in the water.

Nitrate in ground waters can contribute to the nutrient load, and hence eutrophication of surface waters, since ground water may become surface water due to 'inter-

* Contribution from the Department of Soil Science, University of Wisconsin, Madison, 53706, and published with the approval of the Director, Research Division of the College of Agricultural and Life Sciences. Supported in part by a grant from the Tennessee Valley Authority.

flow'. Brezonik and Lee (1968) have estimated that ground water nitrate contributes 50% of the N in Lake Mendota sediments and waters. Little data are available for streams, but the fact that nitrate often occurs in streams during winter months when runoff is negligible indicates that ground water nitrate is a factor in stream pollution (Feth, 1966).

3. Sources of Nitrogen

The amount of N required to bring a soil percolate to the 10 ppm 'critical' NO_3-N level is 2.27 lbs per acre-inch. If, e.g., deep percolation out of the root zone is 6 inches per year, only 13.6 lbs/acre of N as nitrate would be required to give the critical concentration. Thus we are dealing with a problem which must account for a small fraction of the total available N in an agricultural situation.

Nitrogen entering the soil undergoes various transformations depending upon the form of N added and the physical, chemical and environmental factors operating in the soil.

Inputs of N can be grouped under precipitation, organic wastes and plant debris, fertilizers, and N fixed biologically, with the tacit understanding that N in soil organic matter must also be considered as a potential source of nitrate in percolates.

Corey *et al.* (1967) have estimated the sources of available N in Wisconsin cultivated soils (Table I). These data would differ among states as well as regionally and

TABLE I

Sources and estimated amounts of available N in cultivated soils in Wisconsin (1965)

Source	Available N	
	lbs/A	%
Fertilizer	10	9
Legumes	12	10
Precipitation	8	7
O.M. decomposition	45	38
Manure	42	36
Total	117	100

locally within the state. However, they indicate the relatively small input of fertilizer N in relation to native soil N (O.M. decomposition) and animal waste disposal. Nationwide, about half of the N removed by agricultural crops is supplied by fertilizer N (Keeney, 1969).

Improved technology and increased competition have dramatically lowered N fertilizer prices recently, and the amount of fertilizer N used has increased proportionally. In view of the 'critical' 2.3 lbs/acre inch of percolate value, contributions of fertilizer N to the nitrate levels in certain shallow ground water aquifers may be greater than found in some investigations (e.g., Smith, 1967; Stout and Burau, 1967).

Disposal of organic wastes (sewage, manure) has been implicated as a causal factor of high ground water nitrates (Smith, 1967; Stout and Burau, 1967; Stewart *et al.*, 1968; Olsen, 1969). Disposal of high concentrations of nitrogenous wastes in situations where aerobic conditions occur in at least part of the profile and provisions for adequate removal of nitrate by plant growth are not made would almost certainly lead to zones of high nitrate waters. In some studies (Stewart *et al.*, 1968; Olsen, 1969) ammonium and nitrite have also been found in relatively high levels under feedlots, and Hutchinson and Viets (1969) have reported high amounts of ammonia absorption by surface waters near cattle feedlots.

While the absolute amounts of nitrogen in precipitation are not large (Feth, 1966), the seasonal fluctuations and relative yearly constancy of precipitation nitrogen indicates that this source must be evaluated, particularly if undisturbed ecosystems are included in a watershed evaluation.

Biological N fixation occurs through two pathways; fixation of N by free-living soil bacteria and algae and by bacteria living in symbiosis with higher plants (Nutman, 1965; Jensen, 1965). The best estimates indicate that the contribution to the N balance by free-living microorganisms is of little significance agronomically. Relatively large amounts of N (10–200 pounds of N per acre per year) may be fixed by the genus *Rhizobium* living in association with leguminous plants. N fixation by bacteria living with non-legumes (predominately angiosperms) has been found to be an important source of N for these plants. The importance of N fixation in contributing to the N status of undisturbed soils has been pointed out by Stout and Burau (1967). Because biologically fixed N is largely retained in plants, transformations of these forms of N is viewed much the same as organic wastes; only the estimates of the amount of N fixed are difficult.

Many articles stressing N pollution of natural waters have tended to overlook the contribution from soil organic matter. If a soil contains 0.2% organic N and 1–3% per year is mineralized (Bremner, 1965) from 20 to 60 lbs of N/acre is released. This release is stimulated markedly by cultivation, and cultivation of soils has been implicated as providing the bulk of the nitrate in some soil profiles (Stout and Burau, 1967; Stewart *et al.*, 1968). Considering the slow movement of percolate waters in many areas, the high nitrate levels in some aquifers may simply be the reflection of the onset of farming a century or more ago.

The comprehensive review by Feth (1966) indicates that geologic sources of nitrate have also largely been ignored. Nitrates occur in cave, caliche and playa deposits (see also Smith, 1967), and most rocks contain N. In fact, nearly 98% of the world's N is contained in fundamental rocks (Stevenson, 1965). Carbonate rocks also contain nitrate (Feth, 1966). Analysis of a large number of limestone samples from Wisconsin (Chalk and Keeney, 1969, unpublished data) has shown that they may contain up to 15 ppm nitrate. Feth (1966) pointed out the fact that limestone terranes seem to be favored habitats of water high in nitrate, and it would appear that geologic contributions of nitrate should be considered when evaluating sources of nitrate in ground waters.

4. The Nitrogen Cycle and the Nitrate Budget of the Soil Profile

In soils, the predominant storage reservoir of N is soil organic matter (Figure 1). Breakdown of plant and animal debris during microbiological action, and immobilization of inorganic forms of N in living microbial tissue continually adds to soil organic N, while microbial breakdown of organic N compounds continually depletes

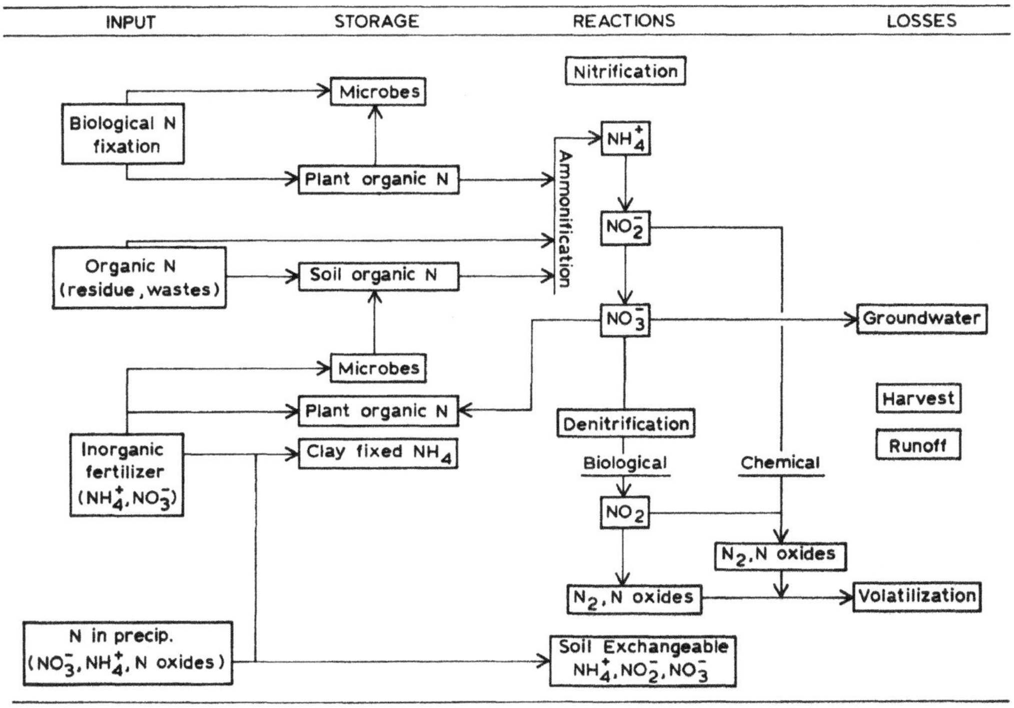

Fig. 1. Detailed nitrogen cycle.

soil organic N (Bartholomew, 1965). Considering the dynamic nature of these processes, the relative stability of soil organic N is quite remarkable. The reasons for organic N stability are numerous, and include formation of resistant heterocyclic N compounds, clay-organic matter complexes, and lack of sufficient carbonaceous energy material for complete breakdown (Bremner, 1967).

The amount of N bound in living microbial cells at a given time is usually estimated at less than 1% of the total N (Bremner, 1967). Clay fixed NH_4, which is relatively unavailable for plant uptake or microbial oxidation, can constitute from 1 to 25% of the total soil N (Keeney and Bremner, 1966; Bremner, 1967; Nommik, 1965).

Ammonification of organic forms of N by heterotrophs occurs under a wide range of pH, temperature and moisture conditions (Bartholomew, 1965; Campbell and Lees, 1967). In contrast, autotrophic nitrification ($NH_4 \rightarrow NO_2 \rightarrow NO_3$) is easily inhibited by unfavorable conditions (Alexander, 1965). Several heterotrophic nitrifiers have been identified, but their importance in nitrification in soils is not known (Alexander, 1965). However, heterotrophic nitrifiers may be important in nitrate formation in organic-rich environments such as composts and manure piles.

The environmental variables affecting nitrification rates are discussed in several reviews (Quastel and Scholefield, 1960; Alexander, 1965; Frederick and Broadbent, 1966; Campbell and Lees, 1967). Briefly, optimal nitrification occurs at 30–35 °C, pH near 7, oxygen concentrations near 20% and at about two-thirds water saturation. Formation of nitrite by *Nitrosomonas spp.* can be inhibited by a large number of compounds (Alexander, 1965). Oxidation of nitrite to nitrate (*Nitrobacter spp.*) is inhibited by high pH's and free NH_3 giving rise to phytotoxic accumulation of nitrite in some alkaline soils treated with anhydrous NH_3 or urea (Alexander, 1965).

Denitrification (reduction of nitrite or nitrate to gaseous forms of N which then can be lost to the atmosphere) can occur by two distinct pathways (Figure 1). Biological reduction of nitrate by soil heterotrophs occurs when insufficient oxygen is available and the organism thus must use another electron source. In most cases, carbonaceous material for energy must be present, which is the reason significant denitrification does not occur in ground water aquifers. Denitrification by chemical reactions occurs only with nitrite (Broadbent and Clark, 1965). Nitrite is unstable in acid soils, giving rise to nitrate and N oxides simultaneously. Nitrite also reacts with soil constituents with the fixation of N in organic matter and gaseous N oxide formation (Broadbent and Clark, 1965; Bremner and Nelson, 1968). The importance of denitrification reactions in soils is just beginning to be appreciated (Wullstein, 1967) and this difficult aspect of the N cycle must be more fully understood and evaluated before predictions on N balance sheets, biological nitrate removal from soils, and N fertilizer rates can be predicted accurately (Allison, 1965).

Addition of carbonaceous material to soils stimulates microbial growth. The N requirements of the microbes are met from the inorganic N pool in the soil (Bartholomew, 1965). These microbes of course decompose, and become part of the soil organic N.

N is the most abundant of the nutrients in plants that must come from the soil or fertilizers (Viets, 1965). Most crops take up N roughly in proportion to their yield, and as such effectively immobilize N from leaching loss, as evidenced by the fact that fallow soils often have higher profile nitrate contents than cropped soils (Stewart *et al.*, 1968; Olsen, 1969). Many agronomic crops effectively remove nitrate from several feet of the profile, and use of deep rooted crops such as alfalfa in rotation with shallow rooted crops has been advocated to reduce nitrate levels in soils (Stewart *et al.*, 1968).

5. Nitrate Leaching from the Soil Profile and Transport to Surface or Ground Water

The concentration of nitrate in the soil solution as it leaves the root zone is not necessarily the same as that which it will have when it reaches the point in the ground water or stream where the concentration is of concern. This is because of the mixing which goes on during the flow process. This mixing, or dispersion as it is called, has been studied in great detail theoretically (Gardner, 1965), but little application has been made to the determination of NO_3 concentrations.

Equations treating the transport of nitrate or other solutions in the soil solution and mixing with ground water are reviewed there. So long as the nitrogen remains in the nitrate form the transport process is quantitatively the same as for chlorides and other salts whose leaching has been studied extensively and is relatively well understood. Uptake of nitrate by plants in the root zone provides an additional complication. However, a knowledge of nitrate concentration in the soil solution with depth in the root zone coupled with a knowledge of the transpiration and deep percolation should provide heretofor unavailable data on uptake mechanisms and the nitrogen balance (Gardner, 1967).

Recent studies on the dispersion phenomenon (Raats and Scotter, 1968) and on the water regime of a transpiring crop (Black *et al.*, 1970) indicate that the necessary tools and concepts exist for a detailed study integrating the major element of the nitrogen and hydrological cycles into a quantitative whole.

6. Research Needs

Interdisciplinary research effort is required to develop the ability to manage the nitrogen cycle through wise use of nitrogen fertilization and waste disposal practices to minimize nitrate accumulation hazards in ground and surface water and enhance environmental quality. This research must develop quantitative mathematical expressions for the principal nitrogen input, transformation and output rates which determine the concentration of nitrate in the soil solution, and combine these models with the equations describing the movement of soil water and solutes through the soil profile to the ground water aquifer or stream. In addition, procedures must be developed so that these models can be applied to solution of actual problems.

The multi-disciplinary research approach discussed in this paper should provide workable and widely applicable models of nitrate movement to ground water, and will go far in answering one of the Nation's pressing pollution problems.

References

Alexander, M.: 1965, 'Nitrification', *Agron.* **10**, 307–343.

Allison, F. E.: 1965, 'Evaluation of Incoming and Outgoing Processes that affect Soil Nitrogen', *Agron.* **10**, 573–606.

Bartholomew, W. V.: 1965, 'Mineralization and Immobilization of Nitrogen in the Decomposition of Plant and Animal Residues', *Agron.* **10**, 285–306.

* Bartholomew, W. V. and Kirkham, D.: 1960, 'Mathematical Descriptions and Interpretations of Culture Induced Soil Nitrogen Changes', *Trans. 7th Intern. Congr. Soil Sci.*, Madison, Wisc., pp. 471–477.

* Biggar, J. W. and Corey, R. B.: 1967, 'Agricultural Drainage and Eutrophication', *Agr. Sci. Rev.* **5** (4), 22–28.

Black, T. A., Gardner, W. R., and Tanner, C. B.: 1970, 'Water Storage and Drainage under a Row Crop on a Sandy Soil', *Agron. J.* (in press).

Bremner, J. M.: 1965, 'Organic Nitrogen in Soils', *Agron.* **10**, 93–149.

Bremner, J. M.: 1967, 'Nitrogenous Compounds', in *Soil Biochemistry* (ed. by A. D. McLaren and G. N. Peterson), Marcel Dekker, New York, pp. 19–66.

Bremner, J. M. and Nelson, D. W.: 1968, 'Chemical Decomposition of Nitrate in Soils', *Intern. 9th Congr. Soil Sci.*, Vol. II, pp. 495–503.

Brezonik, P. L. and Lee, G. F.: 1968, 'Denitrification as a Nitrogen Sink in Lake Mendota, Wis.', *J. Environ. Sci. Technol.* **2**, 120–125.

Broadbent, F. E. and Clark, F.: 1965, 'Denitrification', *Agron.* **10**, 344–359.

Campbell, N. E. R. and Lees, H.: 1967, 'The Nitrogen Cycle', in *Soil Biochemistry* (ed. by A. D. McLaren and G. N. Peterson), Marcel Dekker, New York, pp. 194–215.

* Commoner, B.: 1968, 'Threats to the Integrity of the Nitrogen Cycle: Nitrogen Compounds in Soil, Water, Atmosphere and Precipitation'. Presented at Global Effects of Environmental Pollution Symposium, AAAS, Dallas, Texas, Dec. 26 (see p. 70, this volume).

Corey, R. B., Hasler, A. D., Lee, G. F., Schraufnagel, F. H., and Wirth, T. L.: 1967, 'Excessive Water Fertilization'. Rept., Water Subcommittee, Natural Resources Committee of State Agencies, Madison, Wis. Jan. 31, 54 pp.

* Environmental Pollution Panel, President's Science Advisory Committee: 1965, *Restoring the Quality of our Environment*, U.S. Govt. Printing Office, Washington, D.C.

Feth, J. S.: 1966, Nitrogen Compounds in Natural Water – A Review', *J. Water Res. Research* **2**, 41–58.

Frederick, L. R. and Broadbent, F. E.: 1966, 'Biological Interactions', in *Agricultural Anhydrous Ammonia* (ed. by M. H. McVickar *et al.*), Agricultural Ammonia Inst., Memphis, Tenn., pp. 198–212.

Gardner, W. R.: 1965, 'Movement of Nitrogen in Soil', *Agron.* **10**, 550–572.

Gardner, W. R.: 1967, 'Water Uptake and Salt Distribution Patterns in Saline Soils'. International Atomic Energy Agency Symposium on the Use of Radioisotopes and Radiation Techniques in Soil Physics and Irrigation Studies, Istanbul, pp. 335–341.

Harmsen, G. W. and Kolenbrander, G. J.: 1965, 'Soil Inorganic Nitrogen', *Agron.* **10**, 43–92.

Hutchinson, G. L. and Viets, Jr., F. G.: 1969, 'Nitrogen Enrichment of Surface Water by Absorption of Ammonia Voltalized from Cattle Feedlots', *Science* **166**, 514–515.

Jensen, H. H.: 1965, 'Nonsymbiotic Nitrogen Fixation', *Agron.* **10**, 436–480.

Keeney, D. R.: 1969, 'Nitrate Pollution of Ground Water: How much Blame belongs with Agriculture?', *Fertilizer and Aglime Conf. Abstracts*, pp. 70–75.

Keeney, D. R. and Bremner, J. M.: 1966, 'Determination and Isotope-Ratio Analysis of Different Forms of Nitrogen in Soils. 4: Exchangeable Ammonium, Nitrate, and Nitrite by Direct-Distillation Methods', *Soil Sci. Soc. Amer. Proc.* **30**, 583–587.

* Keeney, D. R. and Bremner, J. M.: 1966, 'Characterization of Mineralizable Nitrogen in Soils', *Soil Sci. Soc. Amer. Proc.* **30**, 714–718.

Mortland, M. M. and Wolcott, A. R.: 1965, 'Sorption of Inorganic Nitrogen Compounds by Soil Materials', *Agron.* **10**, 150–197.

Nommik, H.: 1965, 'Ammonium Fixation and Other Reactions involving a Nonenzymatic Immobilization of Mineral Nitrogen in Soil', *Agron.* **10**, 198–258.

Nutman, P. S.: 1965, 'Symbiotic Nitrogen Fixation', *Agron.* **10**, 360–383.

Olsen, R. J.: 1969, Effect of Various Factors on Movement of Nitrate Nitrogen in Soil Profiles and on Transformations of Soil Nitrogen. Ph.D. thesis, Department of Soils, Univ. of Wisconsin, Madison, Wisc.

Quastel, J. H. and Scholefield, P. G.: 1951, 'Biochemistry of Nitrification in Soil', *Bacteriol. Rev.* **15**, 1–53.

Raats, P. A. C. and Scotter, D. R.: 1968, 'Dynamically Similar Motion of two Miscible Constituents in Porous Mediums', *Water Resources Res.* **4**, 561–568.

Scarsbrook, C. E.: 1965, 'Nitrogen Availability', *Agron.* **10**, 481–502.

Smith, G. E.: 1967, 'Fertilizer Nutrients as Contaminants in Water Supplies', in *Agriculture and the Quality of Our Environment* (ed. by N. C. Brady), Amer Assn. Adv. Sci., Wash. D.C., pp. 173–186.

Stevenson, F. J.: 1965, 'Origin and Distribution of Nitrogen in Soil', *Agron.* **10**, 1–42.

Stewart, B. A., Viets, Jr., F. G., and Hutchinson, G. L.: 1968, 'Agriculture's Effect on Nitrate Pollution of Ground Water', *J. Soil Water Conserv.* **23**, 13–15.

Stout, P. R. and Burau, R. G.: 1967, 'The Extent and Significance of Fertilizer Buildup in Soils as Revealed by Vertical Distributions of Nitrogenous Matter between Soils and Underlying Water Reservoirs', in *Agriculture and the Quality of Our Environment* (ed. by N. C. Brady), Amer. Assn. Adv. Sci., Wash., D.C., pp. 283–310.

Viets, F. G.: 1965, 'The Plant's Need for and Use of Nitrogen', *Agron.* **10**, 503–549.

* Vincent, J. M.: 1965, 'Environmental Factors in the Fixation of Nitrogen by the Legume', *Agron.* **10**, 384–435.

* Wadleigh, C. H.: 1968, *Wastes in Relation to Agriculture and Forestry*. USDA, Misc. Pub. No. 1065, U.S. Govt. Printing Office, Washington, D.C.

* Webber, L. R. and Elrick, D. E.: 1967, 'The Soil and Lake Eutrophication', in *Proc. 10th. Conf. on Great Lakes Research*, pp. 404–412.

* Wheeler, E. W.: 1969, 'Fertilizers are not Threatening our Environment', *Farm Chemicals* **132** (7), 45–48.

Wright, M. J. and Davison, K. L.: 1964, 'Nitrate Accumulation in Crops and Nitrate Poisoning in Animals', *Advance. Agron.* **16**, 201–256.

Wullstein, L. H.: 1967, 'Soil Nitrogen Volatilization. A Case for Applied Research', *Agron. Sci. Rev.* **5** (2), 8–13.

For Further Reading

1. Black, C. A.: *Soil-Plant Relationships*, 2nd edition, John Wiley and Sons, Inc. New York, 1968, pp. 70–152, 405–557.
2. Hine, R. L. (ed.), Water use: Principles and guidelines for planning and management in Wisconsin. Wisconsin Chapter, Soil Conservation Society of America, Madison, Wis., 1969.
3. McVickar, M. H., W. P. Martin, I. E. Miles, and H. H. Tucker (eds.), *Agricultural Anhydrous Ammonia*, American Society of Agronomy, Madison, Wisc., 1966.
4. *Yearbook of Agriculture. Soil*, The United States Dept. of Agr., Washington, D.C., 1957.

* Further references, not specifically mentioned in the text.

NITROGEN COMPOUNDS USED IN CROP PRODUCTION

T. C. BYERLY

Cooperative State Research Service, U.S. Department of Agriculture, Washington, D.C., U.S.A.

Abstract. In order to keep pace with the food needs of the world, application of nitrogen in chemical fertilizer must be and is being increased very rapidly, with a doubling time of about 10 years. The efficiency of use of fertilizer N by crop plants diminishes as rate of application is increased. The N not used by the growing plants may be stored in the rhizosphere, beneath it, percolate to groundwater, or principally, be volatilized to the atmosphere. Efficiency of crop plants in use of N and in its accumulation as nitrate varies widely. Genetic variation is notable; e.g., the chenopods – beets, spinach, and the like – are notable accumulators. Under drought conditions, oats, corn, and other crop plants may also accumulate NO_2. Nitrate in well-water in some areas is presently high; unrelated to fertilizer use.

Research to provide systems for more efficient use of fertilizer N is needed both on grounds of economy and to minimize the accumulation of N from fertilizer in our waters and, as nitrate, in our feed, forage and food plants.

Commoner raises four principal issues. He asserts that:

(1) Nitrate from mineral fertilizer is percolating to groundwater and eventually reaching our lakes and streams and that this nitrate contributes substantially to eutrophication of lakes and streams.

(2) Nitrate content of some vegetables presents a threat to health of young infants.

(3) Nitrate content of drinking water supplies exceeds or approaches tolerance levels in substantial areas of the United States.

(4) Principal dependence for the nitrogen requirements of food crops on nitrogen bound to organic matter would lessen these hazards. With respect to the first issue, surely some nitrate is reaching our waters from mineral fertilizer. More nitrate is likely to reach the environment as nitrogen fertilizer use is increased. At present the principal causes of eutrophication are phosphorus, largely from detergents, and nitrogen from human, plant, and animal wastes.

With respect to nitrate in food plants, nitrate from synthetic fertilizer is only a contributing factor, the principal factors being genetic tendency of chenopods to accumulate nitrate and environmental stresses which accentuate accumulation.

No cases of infant methemoglobinemia attributed to spinach or baby food have been reported in the U.S. It is well known that spinach and beets are nitrate accumulators and that their nitrate content varies widely. While Dr. Commoner cites research findings indicating association of high nitrate content in spinach with high nitrate fertilization [1], equally high nitrate levels in spinach were reported in 1907 before synthetic nitrogen compound fertilizers were in use as were found in 1967 [2].

The issue of nitrate in drinking water existed before commercial fertilizer came into general use. Wherever nitrate content of drinking water exceeds or approaches the public health limit tolerance, an improved drinking water supply should be obtained, regardless of origin of the nitrate. Applications of known technology or new technology which may be developed is an obvious alternate to a new supply source for bringing nitrate content of drinking water supplies within acceptable limits.

The fourth issue, that of maximal dependence on nitrogen bound to soil organic matter to supply nutrient requirements of crop plants, is difficult. First, disturbance of soils high in organic matter by tillage releases large amounts of nitrate in the environment. Second, very few soils are likely now or ever to develop needed reservoir capacity of organic-bound nitrogen to support continuous cropping with the high yields now required for food crops.

I do think we have a problem. That problem should be defined. We need to control the entry of nitrates from commercial fertilizers, as well as from sewage, animal wastes, and other sources, into groundwater, lakes, and streams. The degree of control compatible with essential crop production will depend on the improvement of technology through research and the application of that improved technology. This we can and should do. I do not share Dr. Commoner's alarm. I see no clear and present danger.

We must increase greatly our use of synthetic nitrogen fertilizers in order to feed the world. The world's population will double by the year 2000 despite pills, intrauterine devices (IUD's), or any other restraint. Most of that increase will take place in the developing countries. And in those countries food production must and can be increased, largely on land now under cultivation [3].

With present technology, world use of commercial nitrogen fertilizers may need to be increased by 10 times over current usage. Nitrogen is not the only factor limiting crop yields; other mineral nutrients, water, pesticides, power machinery, and fuel, are all necessary to obtain the doubled crop yields we must have. The new rice and wheat and corn varieties will give double present yields but only with adequate inputs, and nitrogen is essential.

World use of nitrogen in manufactured fertilizer increased from 4272000 metric tons annually in the 1948–52 period to 19844000 metric tons in the 1966–67 period. U.S. use increased from about 1173000 metric tons to 5487000 tons – an increase of more than 4.5 times for the world and for the U.S. [4].

A substantial portion of the fertilizer nitrogen was used on cereal crops such as corn, wheat, and rice. Production of all cereals increased about 60% during the period. Cereal production should be more than double present production in year 2000.

It seems obvious that the large increase in fertilizer nitrogen use which the world must have will lead to some increase in nitrogen in our waters and in our air from this source unless research produces more efficient, economic, technology for fertilizer nitrogen use. Present technology may give us a bushel of soft, white winter wheat with about 10% protein or 1.0 lb. nitrogen per bushel for each 3 lb. of fertilizer nitrogen used [5]. Thus, 1 lb. of nitrogen applied is in the grain; two go to straw, the rhizosphere, the percolate, the atmosphere. Bread wheat in Decatur County, Kansas, returned about 11 bushels more per acre and about 20 lb. more nitrogen in the grain harvested from plots receiving 75 lb. nitrogen per acre than plots receiving no fertilizer nitrogen [6]. And the law of diminishing increments is applicable. In general, the higher the rate of application, the lower the efficiency of nitrogen use [7].

We can double our current corn yields and they should be doubled by year 2000. With our existing technology it may require 300 lb. of nitrogen per acre from chemical fertilizer. One Illinois farmer produced 189 bushels per acre in 1965 on a substantial acreage. He used from 265–337 lb. of nitrogen [8].

Most plants absorb most of their nitrogen from the soil as nitrate. Soil micro-organisms release nitrogen from organic matter in the soil and convert it to nitrate. Bacteria convert inorganic nitrogen compounds in commercial fertilizer applied to the soil to nitrate. In the presence of sufficient moisture, some nitrate may be leached to groundwater, thence to lakes or streams. This is true both of nitrogen derived from organic matter and for that derived from mineral fertilizer. The amount leached is a complex function of several factors including available nitrate, soil moisture, time of year, temperature, nature of vegetation, and soil composition.

Biological efficiency of use of nitrogen requires that a sufficient supply of available nitrogen be present when needed. Thus, winter wheat yields may be increased by nitrogen applications in fall or early spring. Protein content of the grain will be increased only by nitrogen available late in the growing season [6].

Economic efficiency may be highest when nitrogen fertilizer is applied in a single application at the time most convenient to the farmer. This is likely to be fall or spring, outside the growing season. The greater biological efficiency achieved through application as needed by the crop, in most cases, would increase crop production costs and, thus, food costs because of the added costs of fertilizer application.

A great deal is currently being done and more is planned to maintain and enhance the quality of the environment. The role of nitrates in eutrophication is an important, but not the dominant, factor in this problem.

That the people of the world can have an environment of good quality, green with grass and trees, is true because the development and application of improved technology for food production on farms continues to increase output per acre. Increased use of manufactured nitrogen compounds is an essential ingredient of current and prospective farm production technology.

Crops are harvested from less than 300 million acres in the U.S. currently compared to about 360 million acres in 1930. In the world at large and in the United States only modest expansions of harvested acreage will be necessary to feed the world. The FAO Director-General [9] pointed out that, "By and large the developed countries already have more land available for agriculture than they really need". And land not needed for cultivated crops should grow grass and trees. Dr. Commoner noted decrease in nitrates in the waters of the upper Missouri and suggested that improved soil conservation practices may be a factor. I agree that this may be true. The Great Plains Conservation Program, administered by the Soil Conservation Service, has provided comprehensive conservation treatment for more than 50 million acres in the Great Plains, much of it in the upper Missouri drainage [10].

In December 1968 the United States joined with 49 other nations in a resolution to convoke, not later than 1972, an International Conference on the Problems of the Human Environment. This resolution constituted agenda item 91 of the 23rd Sessio of the UN General Assembly [11]. It was adopted without objection. The resolution requested the Secretary-General to report to the 24th Session of the General Assembly the main problems which might with particular advantage be considered at such a conference.

Speaking in support of the resolution, the representative of Sweden, Mr. Aström, said: "There is no longer any hope of finding new important soils to feed an increasing population. It is rather a question of making already cultivated soils yield richer harvests through higher productivity. Such efforts imply the extensive use of fertilizers, pesticides, the building of huge irrigation systems, and so on." And again, "Substances which in the right place are valuable resources may in another place be harmful. An example of that is the spreading of chemicals as fertilizers on the fields. If these chemicals reach lakes and streams, they become pollutants."

Mr. Aström also noted the value of relevant programs for improvement of the environment of the UN Specialized Agencies, the Biosphere Conference held at UNESCO House in September 1968, of the International Council for the Conservation of Nature and Natural Resources, the International Council of Scientific Unions, and the International Biological Program which it sponsors.

U.S. Ambassador Wiggins also spoke in support of the resolution [12].

The U.S. Department of Agriculture and the cooperating State agricultural experiment stations are currently devoting about 220 scientific man-years (SMY) (full-time equivalent of 220 principal investigators and their assistants and graduate students) to research on plant nutrients. The joint task force on environmental quality [13] estimated need for 55 additional SMY's to do research on plant nutrients by 1977. Ten of these additional SMY's would do research related to control of nitrates reaching our waters from agricultural and other rural sources.

The task force estimated (*loc. cit.* p. 62) that: "Improved technical information for different soils, under different climates for different crops could avert as much as one-half of the loss from non-beneficial use of this fertilizer."

Even with present technology, ways are known to increase efficiency of nitrogen use. But they are more expensive than current methods and cannot be justified economically at present. People may have to make some choices between food costs and esthetics.

Research is underway in our country and many others on the metabolism of plants. Nitrogen is supplied by biological processes – blue-green algae and free living bacteria as well as those symbiotic with legumes. Legumes are important as food crops. The IBP, in which scientists in many countries, including the U.S., are participating includes major programs on the factors affecting the metabolism of nitrogen by plants in production of proteins [14].

In summary, we have reason, in my opinion, to be alert, not to be alarmed, with respect to the role of fertilizer nitrogen in eutrophication and in the accumulation of nitrates in food and feed plants. Increased use of commercial nitrogen fertilizer is necessary. The benefits of such use in feeding the people of the world are unquestionable. More research to produce and adopt more efficient technology for use of commercial nitrogen fertilizer and the application of such technology is needed. More research is needed on the biology and ecology of fixation and metabolism of nitrogen by blue-green algae, free living bacteria, symbiotic bacteria and by other organisms. Much more research is needed on nitrogen in all its roles in all our ecosystems and their effect on human welfare.

References

[1] Schupan, von, W.: 1965, 'Der Nitratgehalt von Spinat (Spinacea oleracea L.) in Beziehung zur Methämoglobinämie der Säuglinge', *Z. Ernährungswiss.* **5**, 207.

[2] Richardson, W. D.: 1907, *J. Amer. Chem. Soc.* **29**, 1747.
Jackson, W. A., Steel, J. S., and Boswell, V. R.: 1967, 'Nitrates in Edible Vegetable Products', *Amer. Soc. Hort. Sci.* **80**, 349.

[3] Boerma, A. H.: 1968, 'Food Requirements and Production Possibilities', *Sci. Bios. Inf.* **18**, UNESCO, Paris.
Byerly, T. C.: 1967, 'Benefits and Usefulness of Food Chemicals in Food Production Relative to World Population', in NAS-NRC Publ. 1491, Use of Human Subjects in Safety Evaluation of Food Chemicals', p. 15–29.
Ennis, W. B., Jansen, L. L., Ellis, I. T., and Newson, L. D.: 1967, 'Inputs for Pesticides', in *The World Food Supply*, Vol. III, The White House, Washington, D.C., p. 130–175.

[4] F.A.O.: 1967, *Production Yearbook* **21**, 445. Food and Agriculture Organization, Rome.

[5] Leggett, G. E.: 1959, Relations Between Wheat Yield, Available Moisture, and Available Nitrogen in Eastern Washington Dry Land Areas. Wash. Agr. Expt. St. Bul. 609.

[6] Smith, F. W.: 1964, 'Fertilizing Wheat for Profit', *Plant Food Rev.* **10**, 4–6.

[7] Ennis, W. B., Jansen, L. L., Ellis, I. T., and Newson, L. D.: 1967, 'Inputs for Pesticides', in *The World Food Supply*, Vol. III, The White House, Washington, D.C., p. 130–175.

[8] Strohm, J. C. and Ganschon, C.: 1968, *The Ford 1968 Almanac*, Golden Press, New York, p. 99

[9] Boerma, A. H.: 1968, 'Food Requirements and Production Possibilities', *Sci. Bios. Inf.* **18**, UNESCO, Paris.

[10] Freeman, O. L.: 1967, Resources in Action/Agriculture 2000, U.S. Department of Agriculture, Washington, D.C.
Freeman, O. L.: 1969, Agriculture in Transition, Report of the Secretary of Agriculture for 1968, U.S. Department of Agriculture, Washington, D.C.

[11] UN General Assembly, 23rd Session: 1968, Resolution adopted by the General Assembly. A/Res/2398 (XXIII), December 1968.
1968, Provisional verbatim record of the seventeen hundred and twenty-second meeting. A/PV.1732, December 1968.

[12] Wiggins, J. R.: 1968, Provisional verbatim record of the seventeent hundred thirty-third meeting. A/PV.1733, December 1968.

[13] Byerly, T. C., Acker, D. C., Evans, J. B., Hazen, T. E., Heggestadt, H. E., Schleusener, P. E., Storey, H. C., Hermanson, R., Stubblefield, T. M., Treadway, R. H., Wadleigh, C. H., Yeck, R. G., Brinkley, P. C., Cellman, I., MacKenzie, D., Buckley, J. L., Deevey, Jr., E. S., King, D. R., Porter, R., Bullard, W. E., Geyer, H. G., and Ward, D. J.: 1968, Environmental Quality: Pollution in Relation to Agriculture. Report of a Joint Task Force of the U.S. Department of Agriculture and the State Universities and Land-Grant Colleges.

[14] NAS-NRC: 1968, Man's Survival in a Changing World. U.S. Participation in the International Biological Program. National Academy of Sciences-National Research Council, Washington, D.C.

For Further Reading

1. Nyle C. Brady (ed.), *Agriculture and the Quality of Our Environment*. A Symposium Presented at the 133rd Meeting of the American Association for the Advancement of Science, December 1966.
2. Wastes in Relation to Agriculture and Forestry. Cecil H. Wadleigh, Director, Soil & Water Conservation Research Division, Agricultural Research Service, U.S. Department of Agriculture. Miscellaneous Publication No. 1065, USDA, March 1968.
3. Clifford M. Hardin (ed.), *Overcoming World Hunger*. The American Assembly, Columbia University. Published by Prentice-Hall International, Inc., Englewood Cliffs, N. J., 1969.
4. Cleaning Our Environment: The Chemical Basis for Action. A Report by the Subcommittee on Environmental Improvement, Committee on Chemistry and Public Affairs, American Chemical Society, 1969.

MAN-INDUCED EUTROPHICATION OF LAKES

ARTHUR D. HASLER
University of Wisconsin, Laboratory of Limnology, Madison, Wisc., U.S.A.

1. Introduction

Many lakes the world over are becoming less desirable places on which to live because of nutrient wastes pouring into them from a man-changed environment. Man's activities, which introduce excess nutrients to lakes, streams, and estuaries are rapidly accelerating the process of cultural eutrophication. Excessive enrichment, brought about by population and industrial growth, intensified agriculture, river-basin development, recreational use of public waters, and domestic and industrial exploitation of shore properties, accelerates the deterioration of waters. The process causes changes in plant and animal life which usually interfere with multiple uses of waters, reduce their aesthetic qualities and economic value, and threaten the destruction of precious water resources. Overwhelming excessive scums of blue-green algae and aquatic plants chokes the open water, makes the water turbid and nonpotable. They die, rot and repel human residents with repugnant odors. Organic matter from this crop sinks and consumes the deep-water oxygen vital for fish and other animal life.

Under natural conditions lakes proceed toward geological extinction at varying rates through eutrophication or bog formation. Many lakes, in unpopulated temperate zones, and lying in sandy granite drainage basins are still pristine and clear (oligotrophic) even though 10000 years have elapsed since the glacier formed them. Other lakes in the same area, such as shallow bog lakes which were likewise shaped by grinding ice during the same glacial epoch, are already extinct. They are grown over with mats of sphagnum moss interspersed with orchids and pitcher plants. Brown colored water lies below the mat which deteriorates and slowly fills in the basin. In some, the terminal stages of bog formation are evident because these former lakes are now covered with shrubs, tamarack and black spruce forests. This type of extinction is no teutrophication. How this succession or continuum proceeds from open lake to forest is too complex to be developed in this brief essay.

Archeological studies by G. E. Hutchinson and R. Patrick of cores of lake sediments of the Italian lake, *Lago di Monterosi*, reveal that the Romans by constructing roads inadvertently increased the nutrient drainage of a landscape by cutting the trees and exposing limestone strata. The erosion from these nutrient richer strata was followed by a eutrophic period in the lake's history as recognized by the kinds of diatoms found in the cores. E. S. Deevey, Yale Univ., also recognized prehistoric changes of climate and rate of eutrophication in Linsley Pond, Conn., which are correlated with the fossils in the cores. He determined the abundance and variety of microscopic organisms, plankton crustacea and insect larvae in different strata.

The rate of eutrophication of lakes in geological time can often be predicted by examining the soil and vegetation of its drainage basin. If the drainage area is large, the vegetation pristine and the soil rich and erodable, the lake water will be rich in algae and fish; if poor, the water will produce little and will retain its clarity because of low algae count and high aesthetic quality.

2. The Algal Community (Phytoplankton)

Algae are microscopic one-celled plants which require, for their growth, the same nutrients as do garden flowers and lawns. If fertilized richly in spring and summer they flourish; if impoverished they grow sparsely. A community of free floating algae is much richer and more diverse in kinds of species than is any garden.

Pure cultures of algae grown in flasks and chemostats in the laboratory, multiply and grow rapidly if nitrogen and phosphorus bearing chemicals are added; the algae in similar cultures grow even more luxuriantly if only small amounts of sewage effluent are added containing the same amounts of nitrogen and phosphorus as in the original above, hence demonstrating that in addition to P and N there are ingredients (probably vitamins and growth hormones) in sewage which promote growth.

A nutrient poor temperate zone lake will be clear, hence one might collect, with a cone shaped, fine meshed, silk net, thousands of miniscule algae cells at depths of 150 feet and more. In a nutrient-rich lake, on the other hand, the high numbers of algae, the food of protozoa, rotifers and waterfleas, will lend a greenish cast to the surface water, restrict the penetration of sunlight and therefore limit photosynthesizing algae as well as rooted aquatic plants to the shallower depths.

The upshot of eutrophicated (nutrient-enriched) conditions has adverse ramifications. When the enriched conditions are owing to man-made effluents, the algae grow so profusely that the waterfleas (the basic food of all larval fishes) cannot consume the algae fast enough to reduce their numbers significantly; hence abnormal amounts die uneaten.

The biological communities of a lake become upset when bacteria are unable to convert dead organic matter into plant and animal food.

Not only is oxygen, in the deep cool water, exhausted by organic products but hydrogen sulfide (rotten egg gas) accumulates to poisonous levels.

The finale in these despoiled depths is the demise of all noble fishes, e.g. whitefish, trout and cisco which demand oxygen rich water depths for life. Moreover, some noble fishes such as cisco spawn in the fall – their eggs must incubate throughout winter, but an enriched lake having lost its oxygen in the deep layers (under the ice) cannot nourish the eggs for hatching in the spring.

3. Sources of Nutrients

Phosphate additions appear to be one of the major factors in pollution of European and North American lakes, although the rate at which nutrients pass through chemical and biological cycles is also important. Sources of plant nutrients are principally from human sewage and industrial wastes, including the phosphate-rich detergents as can be seen from Table I. Drainage from farmland is second in importance as a nutrient source in temperate zones, where farm manure spread on frozen ground in winter is flushed into streams during spring thaws and rains. A shocking statistic which points up the gravity of our contemporary situation is that farm animals in the Midwest alone provide unsewered and untreated excrement which is equivalent to that from a population of 350 million people. Also, it is surprising that substantial quantities of nitrates of combustion engine and smokestack origin augment these sources. City streets also provide a source of phosphates and nitrates that has to be dealt with.

TABLE I

Summary of estimated nitrogen and phosphorus reaching Wisconsin surface waters

Source	N	P	N	P
	Lbs. per year		(% of total)	
Municipal treatment facilities	20000000	7000000	24.5	55.7
Private sewage systems	4800000	280000	5.9	2.2
Industrial wastes*	1500000	100000	1.8	0.8
Rural sources				
Manured lands	8110000	2700000	9.9	21.5
Other cropland	576000	384000	0.7	3.1
Forest land	435000	43500	0.5	0.3
Pasture, woodlot and other lands	540000	360000	0.7	2.9
Ground water	34300000	285000	42.0	2.3
Urban runoff	4450000	1250000	5.5	10.0
Precipitation on water areas	6950000	155000	8.5	1.2
Total	81661000	12557500	100.0	100.0

* Excludes industrial wastes that discharge to municipal systems. Table does not include contributions from aquatic nitrogen fixation, waterfowl, chemical deicers and wetland drainage.

In toto the results of man-induced eutrophication are catastrophic as noted in the case history of Lake Zürich, Switzerland, where all noble, deep-water fishes, which had provided gourmet specimens for generations, disappeared within 20 years after sewage from the surrounding villages was changed from the chic sale type to flush toilets.

The Zürichsee, a lake in the foothills of the Alps, offers a sad example of the effects of sewage effluent. It is composed of two distinct basins, the Obersee (50 m) and the Untersee (141 m), separated only by a narrow passage. In the past five decades the deeper of the two, at one time a decidedly clear and oligotrophic lake, has become strongly eutrophic, owing to urban effluents originating from a group of small communities totaling about 110000 people. The shallower of the two received no major urban drainage and retained its oligotrophic characteristics for a longer period. Thus we have an experimental and reference lake side by side.

4. History of Fishing

Minder (1926) observed that hand in hand with domestic fertilization the Zürichsee changed from a whitefish (coregonid) lake to a coarse fish lake. In fact, the trout, *Salmo salvelinus*, and a whitefish, *Coregonus exignus*, disappeared from the Untersee and are no longer common in the Obersee; restocking has not been successful. An upsurge of cyprinid fishes (minnow-carp family), chiefly *Alburnus lucidus*, has been striking; mass harvesting of this species has been practiced recently, while *Albramis brama* and *Leuciscus rutilus* have also become abundant with the progressive eutrophication.

5. History of Plankton Succession

Minder is convinced that the decided increase of plankton is not an expression of a natural ripening process, but is owing to plant nutrients, principally P and N, from domestic sources. The diatom *Tabellaria fenestra* appeared, explosively in 1896. Two years later occurred an eruption of the blue-green alga, *Oscillatoria rubescens*. The latter had been known from the eutrophic Murtenersee for 70 years, but otherwise had not been recorded elsewhere in Switzerland, except that it was reported from the Baldeggersee in 1894 where it had appeared in spring and winter. It had not been seen in the Zürichsee plankton until the 1896 eruption when it replaced the usually dominant *Fragilaria capucina*. When Minder (1926) studied the lake during 1920–24, *Oscillatoria rubescens* appeared in quantities in the surface plankton, with a maximum in fall and winter. Oscillatoria generally flourished in the deeper water of the lake in summer. In 1936 Minder (1938) observed a red scum of it over most of the lake. An odor of fish oil is frequently noticeable in summer. There were 1.75 gm wet weight of algae per liter, chiefly Oscillatoria, on 5 May 1899.

Further evidence for recent sudden increase in biological productivity can be found in bottom sediment studies. These demonstrate that Tabellaria occurs in only the most recent layers, its appearance coinciding with the period of Schröter's observations. Moreover, the modern layers are laminated, at least in the deeper parts of the lake, and everywhere are darker than the underlying sediment. The darkening and the laminated character of the sediments are especially pronounced from 1896 onward, the date being determined by counting the seasonal laminae.

Minder (1938) cites some comparative plankton analyses on the Untersee and Obersee: first, in no series were the biocoenoses of the two sections identical; second, *Oscillatoria rubescens* was never found in the Obersee and Tabellaria very infrequently; third, quantitatively the entire plankton of the Untersee was vastly richer; and fourth, plankton quantities are greater in the Untersee downstream from the town of Rapperswil where most of the sewage enters. Minder also observed the rotifer *Keratella quadrata* as appearing first in 1900. *Bosmina longirostris* largely replaced *B. coregoni* after 1911. Since 1920 one of the Ulothricales has become common in summer.

6. History of Other Limnological Factors

Chemical analyses which prove that certain elements of domestic sewage origin have accumulated markedly are cited by Minder (1918), who shows there has been a gradual increase of chloride ion over a relatively few decades. Analyses in 1888 showed the water contained 1.3 mg Cl/l; by 1916 it had risen to 4.9 mg/l. The organic matter as measured by loss on ignition also rose from 9 mg/l in 1888 to 20 in 1914. Vollenweider (1968) gives as a rule of thumb 0.2–0.5 gm/m^2/yr of P and 5–10 gm/m^2/yr of N as the levels of these nutrients which are associated with nuisance blooms of algae in European lakes.

Minder (1943a) gives comparisons of the changes in transparency (average of 100 readings, see Table II).

TABLE II

	Maximum disc reading	Minimum disc reading
Before 1910	16.8	3.1
1905–1910	10.0	2.1
1914–1928	10.0	1.4

It is significant also that O_2 values in the deep water have decreased in the last four decades (Minder, 1943b). Midsummer values at 100 m were nearly 100% saturation from 1910–1930; from 1930–42, however, they were as low as 9% saturation but averaged about 50%.

7. Man-Made Lakes

Lakes are more adversely affected by sewage effluent than are flowing streams chiefly because flowing water is not conducive to algae growth although diatoms do grow on bottom stones and large aquatic plants grow profusely if the water is not too swift or turbid. Hence, the diversion of sewage around lakes and into streams is the lesser evil. Nevertheless, while alleviating the lake problem it places an increasing burden upon the stream's biological system and upon the communities downstream which must purify it. In modern times most large streams have man-made dams for impounding the water and whose outflow provides energy for hydro-electric power. If such a reservoir receives sewage, the quality of the water deteriorates as does a lake, and the cost of water purification and odor control rises for downstream communities. There is therefore a finite limit. Permissible levels of impurities will have to decrease, hence the technological improvement to obtain more complete nutrient removal must become more efficient.

8. Limnological Features of Eutrophicated Lakes

Figure 1 shows the data obtained by Edmondson (1968) and his associates during their continuing long-term investigation of Lake Washington at Seattle. The change in chlorophyll concentration in the epilimnion corresponds with the trend of the phosphate accumulation, reduced transparency, and increase in the O_2 deficit and are therefore expressions of racing eutrophication.

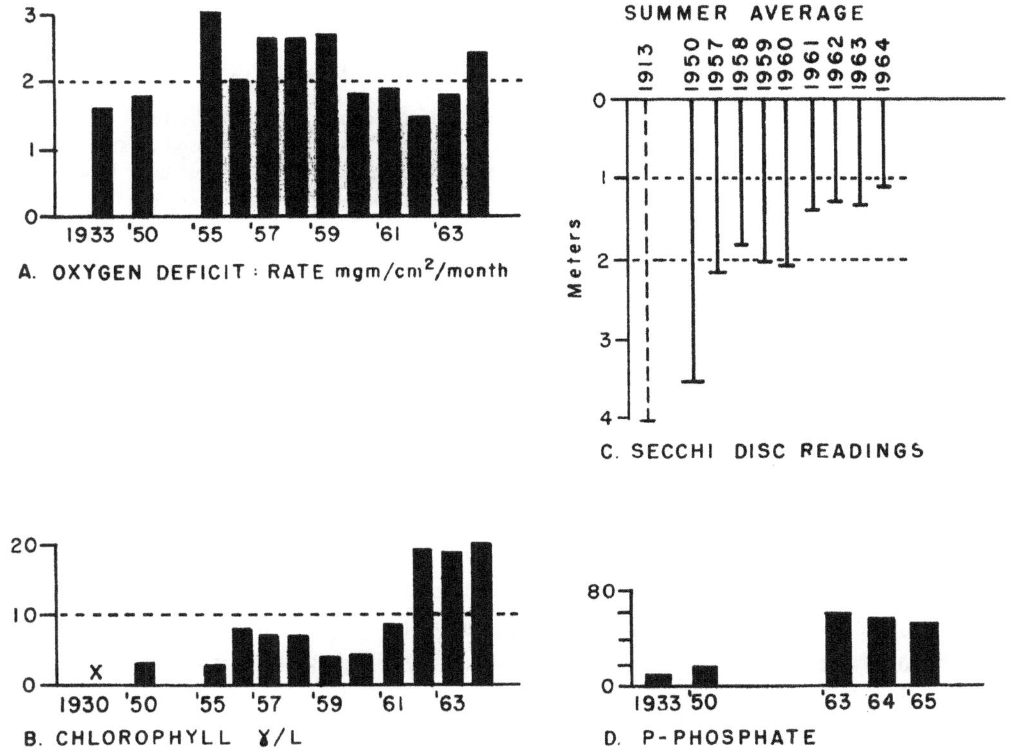

Fig. 1. Changes that have occurred in Lake Washington. (A) Rate of development of relative areal oxygen deficit below 20 m depth between 20 June and 20 September. (B) Mean concentration of chlorophyll in top 10 m of lake during same period of time. (C) Mean Secchi disc transparency, June–September. (D) Maximum concentration of phosphate phosphorus in surface water during winter. (After Edmondson, 1968.)

Findenegg (1964) (Figure 2) has used the ^{14}C method (radioactive carbon) to evaluate degree of eutrophication. It has the advantage that it measures the rate at which energy is converted into carbohydrate in the photosynthetic process. A eutrophic lake fixes carbon principally in the near surface water because its turbidity prevents light from supplying essential energy to algae in the deeper layers. Higher nutrients in a eutrophic lake also serve to stimulate higher rates of ^{14}C fixation in the surface waters.

Dr. Findenegg's example demonstrates a general principle in biology that in senility

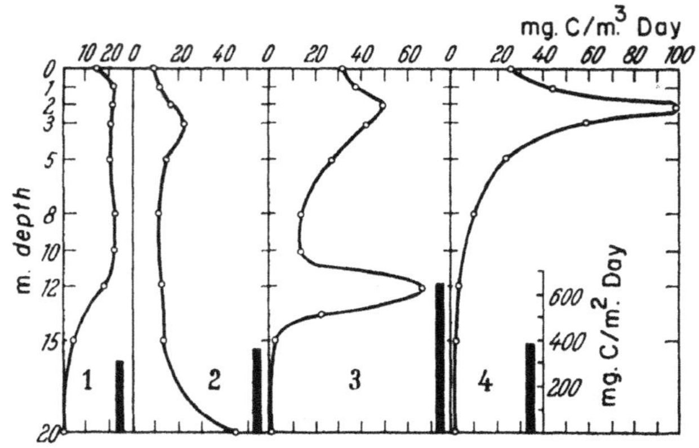

Fig. 2. The production of carbon by photosynthesizing algae in four alpine lakes (1. Millstätter, 2. Klopeiner, 3. Wörther, and 4. lower part of the Lake of Constance) expressed as mg C per m^3 per day. The columns give the total production below 1 m^2 in mg C per m^2 per day. Note the restriction of production to the surface waters in the most eutrophic lake, No. 4.

the older animal often consumes as much food as when he was young, but utilizes it less efficiently. In jargon we say, "He spins his wheels!". The four examples show how with increasing levels of eutrophication there is a steady increase in carbon fixed, but as turbidity rises (4th example), photosynthesis is restricted to the surface waters, hence less depth can be used for production and the utilization drops off – frankly the lake is overfed and obese, perhaps also physiologically senile and – it is spinning its wheels.

Reduction in transparency (Secchi Disc) of Lakes Washington and Zürich appear to be characteristic for rapidly eutrophicated oligotrophic lakes. Hyper-eutrophication of a natural eutrophic lake, Lake Mendota, has not changed the average transparency nor the average hypolimnetic deficit (Stewart, 1965) although other characteristics such as loss of cisco, increase in macrophytes and increase in algae blooms are conspicuous features.

9. Phytoplankton

Comparisons of algae diversity of oligotrophic Trout Lake and eutrophic Lake Mendota show the latter to have fewer species but the size of the organisms is considerably larger indicating higher levels of production than in the oligotrophic lake (Sager, 1967).

Often the low species diversity of the phytoplankton in eutrophic lakes is a result of high populations of blue-green algae such as *Aphanizomenon flos aqua* and *Anabaena spiroides* in Lake Mendota. In some seasons *Fragillaria crotonensis* and *Stephanodiscus astrae* become dominant.

In many eutrophicated northeastern U.S. lakes rooted aquatic plants Myriophyllum and Ceratophyllum become festooned with the filamentous alga *Cladophora* and form dense mats in shallow areas.

10. Great Lakes

Until recently it was thought that eutrophication would not be a major problem in large lakes because of the vast diluting effect of their size. However, evidence is accumulating that indicates eutrophication is occurring in the lower Great Lakes. Furthermore, the undesirable changes in the biota appear to have been initiated in relatively recent years. Charles C. Davis (1964), utilizing long-term records from Lake Erie, has observed both qualitative and quantitative changes in the phytoplankton of that large body of water owing to cultural eutrophication. Total numbers of phytoplankton have increased more than threefold since 1920, while the dominant genera have changed from *Asterionella* and *Synedra* to *Melosira, Fragillaria*, and *Stephanodiscus*.

Other biological changes usually associated with the eutrophication process in small lakes have also been observed in the Great Lakes. Alfred M. Beeton (1965) recently summarized the literature pertaining to the trophic status of the Great Lakes in terms of their biological and physiochemical characteristics and indicated that, of the five lakes, Lake Erie has undergone the most noticeable changes due to eutrophication. In terms of annual harvests commercially valuable species of fish, such as the lake herring or cisco, sauger, walleye, and blue pike, have been raplaced by less desirable species, such as the freshwater drum or sheepshead, carp, and smelt. Similarly, in the organisms living in the bottom sediments of Lake Erie drastic changes in species composition have been observed. Where formerly the mayfly nymph *Hexagenia* was abundant to the extent of 500 organisms per square meter, it presently occurs at levels of 5 and less per square meter. Chironomid midges and tubificid worms now are dominant members in this community.

11. What can be done to Reduce the Galloping Rate of Eutrophication?

The deterioration of our lakes proceeds at such a galloping pace that there is insufficient time to raise an enlightened younger generation which could cope with the causes of eutrophication, hence every effort must be undertaken to convince government officials and voters that action, even though expensive, must be taken immediately to avoid catastrophe. In order to obtain positive returns, time operates negatively against delay. To insure a brighter future, universities, colleges, churches, service clubs, the press, radio and television must acquire a knowledge of the causes, prevention, and cure and begin without delay and help to disseminate factual information.

Provided they are given the facts, preachers and rabbis could preach sin against the environment as convincingly from the pulpit as they preach sin against the soul.

Decision makers, such as legislators, state, county, and village officials, decision planners such as architects, decision formulators such as lawyers and judges, decision executors such as realtors, engineers and contractors must all receive enlightenment about the implication and possible perturbations of the landscape whose environmental health influences the well being of the lake into which the land's effluents flow.

I urge that every educational body, in every community, organize week-long intensive clinics, seminars or working groups to which experienced limnologists and ecologists are invited as teachers, lecturers, demonstrators and guides. I urge journalists and editors, television and radio directors to send their personnel to clinics, or to meet with experienced ecologists in order to obtain facts and illustrations, the basis for which readable and effective articles can be written and programs prepared. They should be taken on field trips to areas where the problem can be demonstrated at first hand in order to demonstrate to the writer or programmer the reality of a eutrophied lake, capture their enthusiasm and stimulate their originality toward the preparation of dynamic and imaginative programs.

The processes of eutrophication are too rapid to risk delay in taking legal action. In applying new concepts of water law to the alleviation of eutrophication there is a need for proper zoning ordinances and forthright public initiative in modernizing the law when the scientific data, even if not complete, suggest action. A new law in Wisconsin requires a 1000-foot setback for all cottages and buildings on lakes and a 300-foot setback on streams, together with stricter specifications for septic tank construction depending upon soil percability.

I would urge legislative lawyers to draft critical legislation for water usage, in regions where the legal procedures are inadequate and encourage them to draft laws which will provide adequate protection of a lake or reservoir from perturbations.

In Wisconsin the late Prof. Jacob Beuscher, through his association with ecologists and landscape architects, drafted unique aforementioned zoning legislation for Wisconsin which now has been passed (Wisconsin Water Resources Law, 1965). Hence, if a dwelling or resort is planned it would have to meet exact specifications for sewage disposal so that no effluent could seep into the water. Some soils are less able to absorb effluents than others, hence the setback of a planned hotel or dwelling must be at the extreme end of 1000 feet if the soil has poor drainage qualities. His legislative bill also specified beauty for this zoned corridor as a quality to be preserved. Vilas County, rich in lakes, prohibits cutting the natural vegetation from more than 10% of the shoreline fronting a property.

12. Social, Legal, and Economic Aspects

In the preceding discussion emphasis was placed upon the effectiveness of various management procedures. Of equal importance are comprehensive economic analyses of new approaches to management. Included should be studies to develop methods to quantify costs and benefits and to analyze public opinion so that the management programs developed are acceptable to society.

Beuscher (1969) writes:

> Since resource management requires not only scientific knowledge and techniques but also governmental and legal structures by which desired management can be achieved, the entire field of legal and governmental structure is a necessary research area. Wisconsin's assertion that it is trustee of all navigable waters of the state is one of the strongest examples of a state's assertion of its right and

duty to protect public interests in natural resources, and this could form a basis for a variety of strong regulatory policies. Potential conflicts between the asserted trusteeship and the rights of private littoral and riparian owners exist, however, and should be investigated as a guide both to the potentials for regulation and to the limitations on regulation without compensation.

Zoning is one type of regulatory action which is likely to be of significant value in attacking the problems associated with inland lakes. As for other water resources, the Wisconsin Legislature has acted, and the counties are presently required by law to enact river and lakeshore zoning ordinances. Research is needed to review the powers of the counties under present statutes, especially noting where powers which seem necessary to accomplish desired regulation are lacking or unclear. Creative proposals and careful analyses are needed concerning the present procedures for administration of zoning by the counties. Projects in various areas of scientific research could be undertaken with a view toward producing facts and testing procedures which the counties might employ to guide and defend their regulation of stream and lakeshore lands.

Owing to the traditional lack of compensation for regulations imposed, zoning is a limited device for the control of lakeshore lands. Imaginative legal research is needed on a broad range of new control devices, such as compensable regulation and partial condemnation, some of which are being tried in some parts of the country. Finally, the powers of all levels of government and the potential powers of private groups and resource control and management corporations should be analyzed as means toward proposing more systematic and creative methods of management than exist presently.

13. Examples of Success

Our knowledge of what causes eutrophication is already sufficiently good that firm and effective precautions can be recommended. They may be expensive to achieve, but the predictive facts are at hand. While improvement in methods can be made more economical, it is not a lack of knowledge which prevents us from action. Three case histories are at hand:

1. Lake Monona

Complaints about the unpleasant odors arising from Madison's Lake Monona were published in the newspapers as early as 1850. Sewage effluents were impugned as the villain in 1885 when a consultant J. Nader advised "... that the lakes were not properly used as receptacles for sewage in the crude state". In 1895 a $1\frac{1}{2}$ mill sewer tax was imposed on assessed valuation, but the sewage treatment plant built from these funds failed in 1898. Septic tanks and cinder filter beds were then constructed but reached capacity in 1906, but it was not until 1914 that a modern sewage treatment plant was constructed. Its effluent entered Lake Monona and its fertility continued to feed the algae and weeds. The process of eutrophication accelerated on Lake Monona and in 1920 its city council minutes read:

> Winds ... drive detached masses of putrefying algae onto shores ... if stirred with a stick, look like human excrement and smell exactly like odors from a foul and neglected pig sty.

In 1921 consideration was given to piping the effluent to the Wisconsin River but another plant was built below Lake Monona in 1928. One half of Madison's effluent then first passed into Lake Waubesa; later (1936) all of it. In spite of heavy applications of the algal poison $CuSO_4$ to the lake to halt the growth of well fed algae, the build-up of offensive and obnoxious odors in Waubesa and Kegonsa worsened.

In desperation the Lewis Anti-Pollution Law was introduced in 1941 but was vetoed by Governor Julius P. Heil because of conflicting issues on whether sewage or rural run-off was the culprit.

In 1942 the Burke Plant which had been discontinued in the '30's was reopened to accommodate the military needs of Truax Field. The need for more copper sulfate during this period is obvious from the graph. In 1943 the Lewis bill was passed to take effect 1 year after the war.

In actuality the effluent did not by-pass the lakes until 1958. During this century of time, buck passing, economy measures, false information from communications, unconclusive action, and lack of cooperation between government and citizens hampered progress.

The fact that copper sulfate treatment of the lake to curb the burgeoning algae growths dropped from freight car load quantities to minimal local treatments is proof that even though agricultural drainage unfortunately continues, the diversion of city sewage produced a change for the better.

2. Lake washington

Lake Washington was used for the disposal of first raw sewage and later treated sewage from the City of Seattle. About 1930, the last major source of raw sewage was removed from the lake, but up until 1959 untreated sewage still entered the lake in relatively small quantities through storm sewer overflows at times of heavy rainfall as well as from seepage from septic tanks whose effluent trickled through or across the ground into small streams entering the lake. In 1959 there were 10 sewage treatment plants, serving 64000 people, putting treated effluent into Lake Washington.

While biologists and engineers warned the community of the impending doom of the lake, it took the dense blooms of *Oscillatoria rubescens*, the same lavender colored alga that produced nuisances in Lake Zürich, to awaken the citizens to the reality of these warnings. While some argued as they did in Madison that the run-off from fertile land was causing the nuisances, others contended that the major sources (city sewage) could be diverted. Radio and television debates were heard and viewed. 5000 women of the League of Women's Voters knocked on doors in the campaign. In addition citizen's groups held meetings culminating in sufficient public opinion to support the city officials in a bonding campaign amounting to an anticipated expenditure of $121000000 for the diversion of 50000000 gallons of sewage around Lake Washington and to empty the sewers into Puget Sound. While the sewers are not yet entirely complete a major part of the sewage has now been diverted. Already the quality of the water has improved noticeably, and the measurements of clarity are improving.

3. Lake tahoe

In order to protect the pristine beauty and crystal clarity of Lake Tahoe several sanitary engineers (McGauhey *et al.*, 1963) made a study of the sewage disposal problems of Lake Tahoe and published a comprehensive study in 1963. They described the problem and projected the rate of growth of the communities and tourist facilities whose sewage from treatment plants and septic tanks is entering this beautiful mountain lake. The lake's great size and immense depth meant that it could absorb some sewage without showing general signs of deterioration yet at the sewage outlets objectionable growths of green algae accumulated on the stones. Hence it was only a matter of time before the increased sewage load from a skyrocketing population would change this clearest of all North American lakes to a lake of lesser esthetic value.

In spite of the 70 odd governmental units in Nevada and California which surround the lake, this initial limnological and engineering evaluation inspired the creation of citizen-governmental action committees and associations which went into action. Already South Tahoe (1968) has built a $19000000 sewage treatment plant in which the treated effluent is pumped over a 7500 ft. pass to a reservoir, over the mountain, where the water will be used for irrigation, and a fertile water at that! Some $10000000 of the total budget was in federal grants acquired by the South Tahoe Public Utility District to help offset this cost – a demonstration of cooperative action of federal and local government in solving a local and national problem. Acknowledging the imminent danger of despoiling this esthetic and financial resource, other communities are facing reality in an action program. A small community of Round Hill in Nevada with only 42 voting citizens, albeit many are owners of gaming houses, has bonded itself for 5.8 mill to treat its sewage and pump it out of the basin. We can only hope that the hotel and residential sewage from other parts of this once gin-clear lake can be similarly diverted in time to avoid the certain despoilation of Sierra Nevada's most magnificent landscape gem.

European lakes, Schliersee and Tegernsee in Germany's Bavarian Alps were eutro-

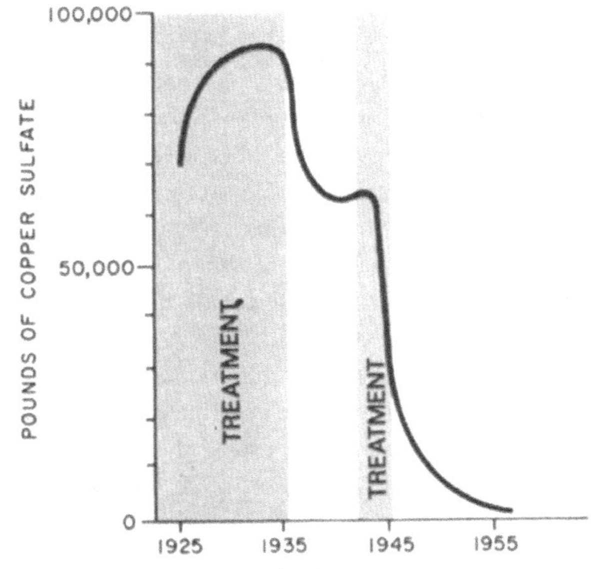

Fig. 3.

phicated by hotel sewage, but are now slowly reverting to more tolerable conditions following a diversion of sewage to the outlets. Lake Ici, France (near Lake Geneva) is following suit. All lakes, from which sewage has been diverted, have shown improvement (see case for Lake Monona, Figure 3) therefore demonstrating that ingredients in sewage contribute greatly to eutrophication and if withdrawn, improvement sets in. These facts negate the argument "Why divert sewage at great cost if rainwater and rural drainage is so rich in nitrogen?" The 'healing' is, of course, more rapid in lakes in oligotrophic and high rainfall landscapes. The Madison, Wis., lakes were naturally eutrophic, nevertheless the hypereutrophic and repulsive conditions do become milder after diversion. In lakes Waubesa and Kegonsa after diversion several species of algae replaced the highly eutrophic single species blooms.

14. Harvesting and Utilization of Excess Crops of Plants and Fish

Machinery, some still in the idea stage, is needed for harvesting large aquatic plants. Removal of this crop from the lake along with its phosphorus and nitrogen containing organic matter impoverishes the water of nutrients. It also improves the esthetics, opens the water area to boating and swimming and creates better shoreline sanitation. More research needs to be done to find a commercial product for aquatic plants and to utilize them for a useful purpose. Eutrophicated lakes produce large crops of fish which should also be harvested more intensively, by commercial fishermen if necessary, because the unharvested fish dying of old age decompose, adding nutrients to the already over-rich environment.

In Lake Mendota in late 1966, 40 lbs./acre of carp alone were harvested in a single seine haul. Yields of 250 lbs./acre/yr. of fish of all species could be harvested easily from this lake without damaging the fishery. In terms of nitrogen and phosphorus 1000 lbs. of rotting fish would yield to the lake 25 lbs. of N and 2 lbs. of P.

15. Chemical Control of Nuisance Growths

Chemicals which poison unwanted aquatic plants and algae have deleterious effects in and around treated areas. Moreover, the killed weeds rot and add to the nutrient supply. This is a bad conservation practice because no good is accomplished. Chemicals distort the structure of multi-species aquatic communities and hence are less useful in lakes than they are in agriculture, where weeds are to be eradicated from a crop of a single species such as wheat. Herbicide use cannot be justified in a lake ecosystem. In addition, chemicals are more difficult to manage than on land for they are soon drifted to other areas. The toxic actions of these chemicals on other species in the lake have not been tested, nor have the possible insidious side effects of sublethal actions over longer periods. At present, the use of chemicals to combat algae blooms or rooted aquatic vegetation can be no more than a palliative. They should be used in aquatic ecosystems only as a last resort. What is eradicated is sure to be replaced by something else that may be more difficult to poison.

16. Utilization of Sewage and Farm Manure

Highly valuable fertilizers are found in sewage (Chicago had 30 tons P per day effluent in 1960). In fact each human produces 1.5 to 4 lbs./yr.

After sewage has undergone secondary treatment it still contains phosphorus. The average P content of sewage (secondary) is 8 mg/liter. Hormones, vitamins and growth substances are also fertilizing ingredients. Moreover phosphate rich detergents now added are not entirely removed. In fact, secondary treatment in most treatment plants removes only about 80% of the P and high costs deter removing more. An increasing human population adds to the total residual left in sewage effluent after treatment. "We seem to be on a treadmill" comments, G. A. Rohlich, an eminent engineer, who states further that in spite of new advances we are not much further ahead of the problem than at the turn of the century.

The price of clean water may rise to a point where we may have to insist upon and want to afford evaporation of the effluents in order to obtain a dry solid and distilled water, as is done in some types of desalting techniques. Secondary treatment of sewage does not remove organic growth factors but probably produces them, hence the evaporation looms as an essential though expensive treatment.

In temperate climates of North America it is customary to scatter farm manure on the frozen land. Large amounts of valuable fertilizer are flushed into streams and lakes during early spring thaws and spring rains. Because a cow produces 6 lbs. nitrogen and 1.5 lbs. phosphorus/yr. it is clear that this source is important. Modernization of the European method of fluidizing dairy cow manure, storing it in huge tanks and distributing it with a 'honey wagon' as soon as the soil can absorb it, is now being recommended. However, economic limitations inhibit progress in converting to a more efficient method of manuring. Fortunately, forest and agricultural soils have a remarkable tenacity for phosphorus. Agricultural and forest crops could profit from the fertilizers from our domestic wastes, but the technology for processing and distributing it are still very expensive when compared with the cheapness of sacked artificial fertilizer.

The volume and weight of dried sewage to be disposed of will have staggering proportions. Settlings from primary treatment of sewage abound near every city, but a farmer can buy and distribute sacked fertilizer cheaper than he can haul dried sewage sludge which is available free of charge. We are too affluent to be able to afford the use of our 'night soil'.

The City of Milwaukee markets in paper bags a dried sludge called Milorganite from its primary settlings which is rich in organic matter. Every city could do this, but instead it piles up and presents a disposal problem because Milwaukee's product satisfies the available market for this product.

17. Benefits of Guided Eutrophication

All eutrophication is not necessarily bad. Well planned enrichment could increase the production of food organisms for fish and hence raise the protein productivity of a natural or man-made lake. Because of the complexity of interactions at various depths and seasons, more knowledge than we now have is needed before we can guide eutrophication and harvest the fish without exceeding the fertility levels that destroy the esthetics of a lake.

18. Predictions

Predicting the consequences of eutrophication would be highly desirable for decision-makers. Systems analysis employs new techniques for constructing mathematical models of a drainage basin to make it possible to evaluate changes which might take place as various eutrophicating factors occur and hence is a powerful tool in dealing with these complex problems in which multifactor cause and effect are involved.

19. In Summary

It is now of greatest urgency to prevent further damage to water resources and to take corrective steps to reverse present damages. Suggested preventive and corrective measures include removing nutrients from municipal, industrial, and agricultural wastes, diversion of treated effluents from lakes, harvesting algae, aquatic plants and fish from lakes in order to help impoverish the water and to improve esthetic qualities; establish regulations for shoreland corridors in order to protect lakes from further damage.

References

Beeton, A. M.: 1965, 'Eutrophication of the St. Lawrence Great Lakes', *Limnology Oceanography* **10**, 240–254.

Beuscher, J.: 1969, in *International Symposium on Eutrophication – Eutrophication: causes, consequences, correctives* (University of Wisconsin, Madison, June, 1967). National Academy of Sciences, National Research Council, Washington, D.C.,

Davis, C. C.: 1964, 'Evidence for Eutrophication of Lake Erie from Phytoplankton Records', *Limnology Oceanography* **9**, 275–283.

Edmondson, W. T.: 1968, 'Water-Quality Management and Lake Eutrophication: The Lake Washington Case', reprinted from *Water Resources Management and Public Policy*, (ed. by Thomas H. Campbell and Robert O. Sylvester), University of Washington Press, Seattle, p. 139–178.

Findenegg, I.: 1964, 'Bestimmung des Trophiegrades von Seen nach der Radiocarbonmethode', *Naturwissenschaften* **51**, 368–369.

McGauhey, P. H., Eliassen, R., Rohlich, G. A., Ludwig, H. F., and Pearson, E. A.: 1963, 'Comprehensive study of protection of water resources of Lake Tahoe', to Lake Tahoe Area Council Engineering-Sciences, Inc., Arcadia, Calif.

Minder, Leo: 1918, 'Zur Hydrophysik des Zürich und Walensees, nebst Beitrag zur Hydrochemie und Hydrobakteriologie des Zürichsees', *Archw. Hydrobiol.* **12**, 122–194.

Minder, Leo: 1926, 'Biologisch-chemische Untersuchungen im Zürichsee', *Rev. Hydrologie* **3**, 1–69.

Minder, Leo: 1938, 'Der Zürichsee als Eutrophierungsphänomen. Summarische Ergebnisse aus fünfzig Jahren Zürichseeforschung', *Geologie Meere Binnengewasser* **2** (2), 284–299.

Minder, Leo: 1943a, *Der Zürichsee im Lichte der Seetypenlehre*, Naturforschenden Gesellschaft in Zürich, 83 pp.

Minder, Leo: 1943b, 'Neuere Untersuchungen über den Sauerstoffgehalt und die Eutrophie des Zürichsees', *Archw. Hydrobiol.* **40** (1), 279–301.

Sager, P. E.: 1967, *Species Diversity and Community Structure in Lacustrine Phytoplankton*, Ph.D. Thesis, University of Wisconsin, 201 pp.
Stewart, K. M.: 1965, *Physical Limnology of some Madison Lakes*, Ph.D. Thesis, University of Wisconsin, 167 pp.
Vollenweider, R.: 1968, personal communication.

For Further Reading

On the causes, consequences, and corrective measures of worldwide eutrophication:

Proceedings of International Symposium on Eutrophication – *Eutrophication: causes, consequences, correctives* (Univ. of Wisconsin, Madison, June, 1967), National Academy of Sciences, National Research Council, Washington, D.C., 1969, 661 pp.

PART III

EFFECTS OF ATMOSPHERIC POLLUTION ON CLIMATE

INTRODUCTION

Human activities are not only increasing the content of carbon dioxide in the atmosphere (see Part I), but also the particle content: dust, smoke, aerosols, even water droplets and ice particles in the high stratosphere, and rocket exhausts in the mesosphere above the stratosphere.

The effects are not at all well understood; even the immediate effects are difficult to predict and our confidence in predicting long-term effects is not high. Yet, not only local weather patterns, but indeed the world's climate, are involved.

Bryson and Wendland attempt to delineate the current thinking and review available scientific studies. They opt for a downward trend in planetary temperatures, over the long run, because of the increased albedo effects of atmospheric particulate material. This trend may accelerate, they believe, if high altitude planes produce a significant increase in the global cirrus cover through contrails.

Mitchell uses a different, but complementary, approach with, again, differing results. He concludes that the carbon dioxide increase is more effective in raising planetary temperatures than is the human-derived particulate loading in reducing temperatures. Natural dust loads, particularly due to volcanic eruptions, may play a more important role.

Manabe examines some of the fundamental assumptions in a critical manner. The induced temperature changes depend on the optical properties of the particles. Under certain circumstances, the increased particulate loading could even raise planetary temperatures.

Schaefer describes more local effects on the weather produced by air pollution. He is especially concerned with the sources of small particles that are introduced into the atmosphere; they can provide condensation nuclei for water vapor or lead to the formation of ice crystals, sometimes over very wide areas.

CLIMATIC EFFECTS OF ATMOSPHERIC POLLUTION*

REID A. BRYSON and WAYNE M. WENDLAND
Dept. of Meteorology, University of Wisconsin, Madison, Wisc., U.S.A.

Abstract. The trend of world temperature in this century appears to be directly related to the trends of atmospheric carbon dioxide content and atmospheric turbidity (dustiness). Both are believed by various scholars to be related to human activities. Since 1940, the effect of the rapid rise of atmospheric turbidity appears to have exceeded the effect of rising carbon dioxide, resulting in a rapid downward trend of temperature. There is no indication that these trends will be reversed, and there is some reason to believe that man-made pollution will have an increased effect in the future.

1. World Distribution of Atmospheric Particulates

Most northern hemisphere meteorologists live in North America, Europe, The Far East, and India. Among these, interest in air pollution is largely directed toward the immediate problems of industrial pollution in cities, and indeed this is the most apparent problem. These are also the regions of most dense observation. However, large sections of the world are characterized by at least high seasonal levels of dust and smoke which are general in distribution rather than concentrated in the cities. Surveys of the literature and personal reconnaissance indicate that this area includes much of Africa, Arabia, southern Asia (especially northwestern India and West Pakistan), China, and Brazil. Of these areas Brazil, Southeast Asia, and Central Africa have a blue haze, probably smoke from agricultural burning, and the rest a brown haze of mineral aerosol (Figure 1). Little attention has been paid to the climatic effect of these millions of square miles of 'dust cloud' cover, the ultimate fate of the dust when advected out of the region, or the general quantity of material resident in the atmosphere. In some of these dusty regions it would appear that the high pollution level is sufficiently constant that it does not attract the attention of the resident meteorologists as a current problem.

The Harmattan haze of Africa, the summer dust storms of India, and the dust flowing out of China on the winter monsoon are well known in descriptive terms, but there is little quantitative information on the densities and depths of the dust clouds. Field observations in India indicate that common densities are 600–800 $\mu gm/m^3$ up to a height of 3000 to 9000 m [1]. Comparison of visibilities suggests that over much of the brown dust areas densities similar to those over India are common. Densities are not known in the blue haze areas, though the low general visibilities of less than three miles often observed during the dry seasons and observed height of the smoky layer reaching in excess of 5000 m indicates that the densities are not negligible.

* This work was sponsored by National Science Foundation Grant GP-5572X1.

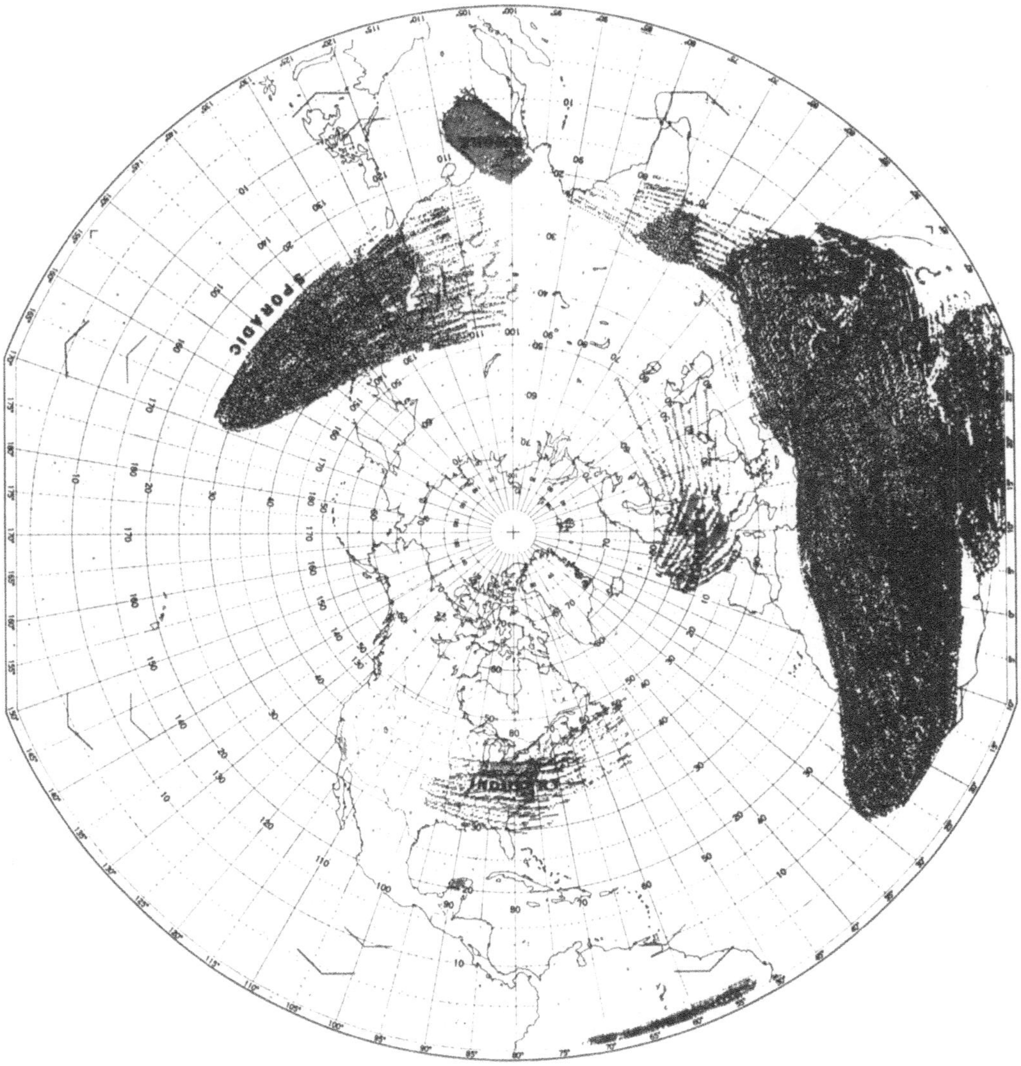

Fig. 1. Schematic distribution of non-aqueous aerosols in the northern hemisphere.

2. Sources

It is generally assumed that the brown dust consists of clay and silt sized particles deflated from the desert surfaces, i.e. is natural [1] while the blue haze is due to slash-and-burn agriculture. There are indications, however, that even in the brown dust areas the rate of deflation and the quantity of dust resident in the atmosphere is strongly influenced by human activities. With atmospheric mixing, this implies that the total worldwide particulate load, adding the usual particulate pollution of the industrialized nations, is subject to human influence of considerable magnitude. Qualitative evidences of this human contribution are to be found in the denser dust streams crossing the Atlantic to the Caribbean after the North African tank battles of World War II had disturbed the 'desert pavement'.

Since DDT is a very, very small fraction of the pollution put into the environment by man, the fact that as much as 14 parts per billion of the dust that falls on Barbados is DDT (DDE) clearly indicates a considerable total human contribution to the dust fall [2]! Even clearer evidence is seen in the parallelism of dust-fall in the Caucasus measured by Davitaia [3] and the capital input to the Soviet economy (Figure 2). Davitaia has expressed the belief that the trend of his dust-fall measurements is due to mechanization and industrialization of eastern Europe [4].

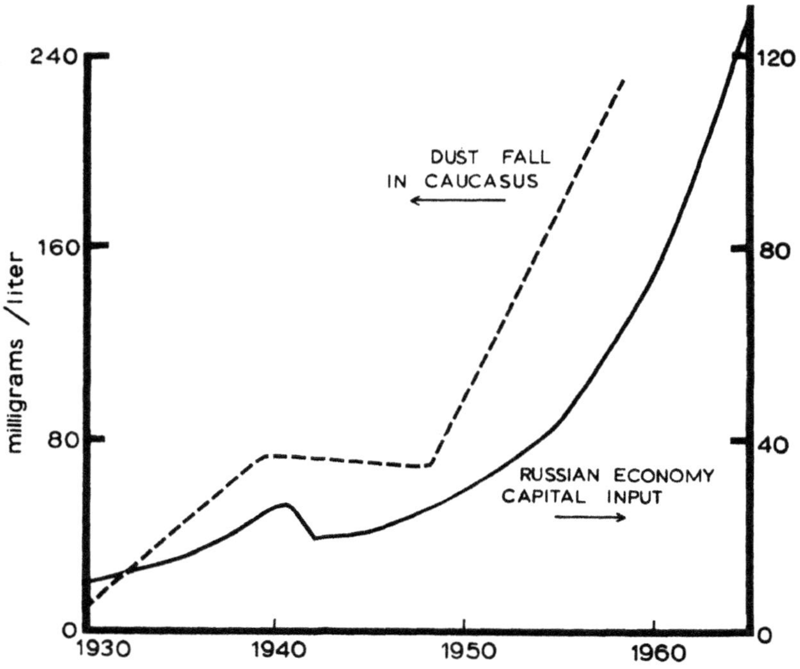

Fig. 2. Dust fall in the high Caucasus [3], and capital input into the Russian economy [22].

3. Trends

It is from the consideration of such trends of dustiness or turbidity that we may investigate the relation of man, dust, and climate. There are far too few studies of long term trends, but most studies have shown a rapid increase in the period since the 1930's quite in addition to those contributions due to volcanic activity. McCormick and Ludwig [5] have shown that the turbidity of the atmosphere over Washington, D.C. and Davos, Switzerland increased 57% (1905–64) and 88% (1920–58) respectively. Davitaia's study of dust-fall on the high snowfields of the Caucasus indicates very little variation from 1790 to 1930, then a catastrophic rise of 19-fold to 1963 only levelling off during World War II (Figure 3). Peterson and Bryson [6] have shown that, if the published figures from the ESSA observatory on Mauna Loa are correct, there was a 30% increase in turbidity over the Pacific between 1957 and 1967 (Figure 4).

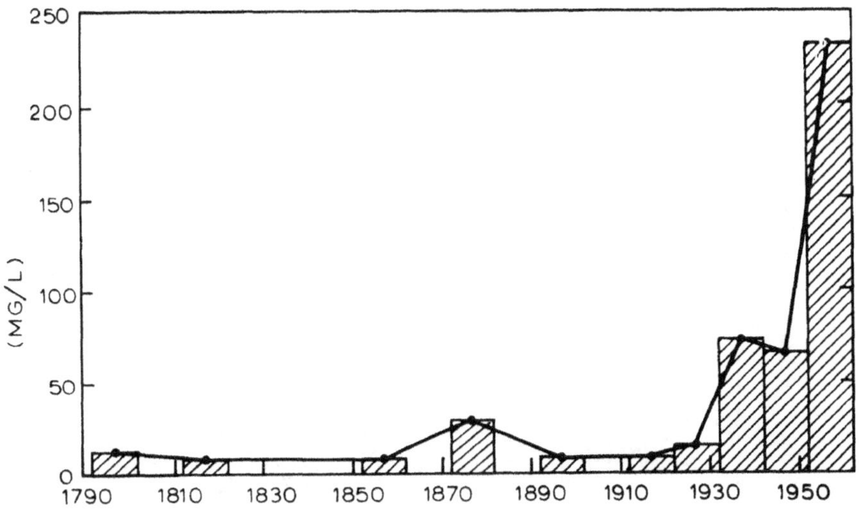

Fig. 3. Dust fall in the high Caucasus [3].

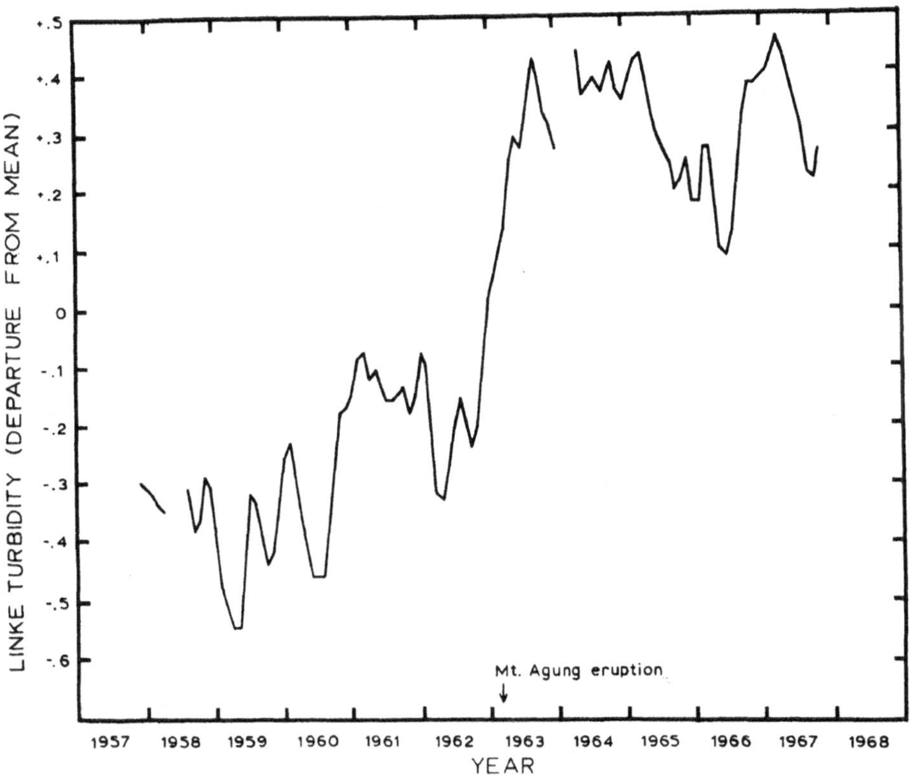

Fig. 4. Trend of monthly anomalies of Linke turbidity factor calculated from published normal incidence radiation data for Mauna Loa, Hawaii. Three-point binomial smooth has been used [6].

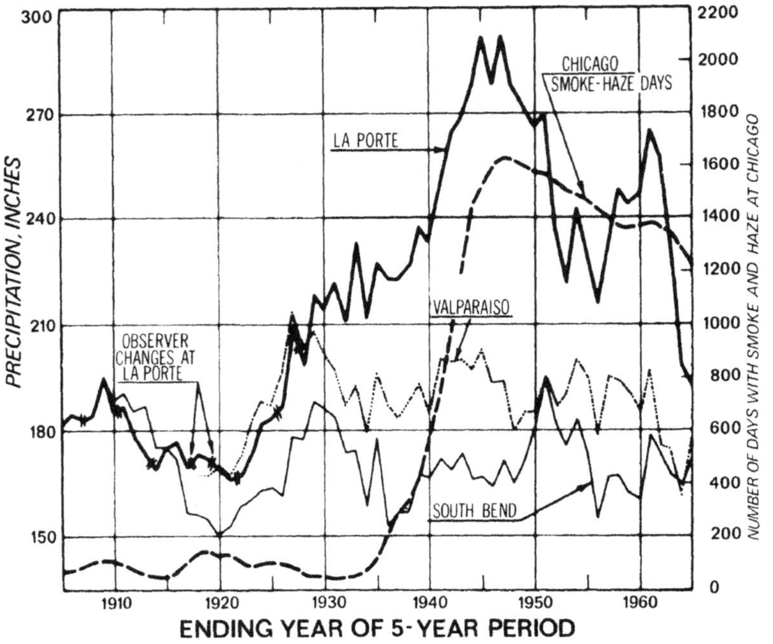

Fig. 5. Number of smoke-haze days per five years at Chicago. Values plotted at end of five year period. [7]

Similar studies of trends over the continent are very sensitive to changes of wind direction. The number of smoky days in Chicago rose from about 20 per year in the decades prior to 1930 to a high of about 320 per year in 1948 [7] (Figure 5). It has decreased somewhat in recent years as the frequency of east winds decreased. It is likely that many industrial areas show the same trend as Chicago.

4. Effects

The local climatic effects of particulate air pollution are better known than the regional and global effects, but far too little is known at any scale. Particulate effluents from industry and energy production apparently seed clouds [7, 8]. Over a city the smoke pall changes the character of the radiation falling on the city [9], especially reducing the ultraviolet penetration, rearranging the long wave radiation balance and providing condensation nuclei which contribute to greater precipitation over a city and a higher frequency of fog [10].

The regional effect of large quantities of suspended dust has been studied in Northwest India [1, 11]. It has been found that the presence of the dense dust in that area affects radiative transfer through the atmosphere such that diabatic cooling of the mid-troposphere by infrared divergence is increased 30 to 50%. Das [12] and Nagatani [13] have shown that the effect of increased diabatic cooling is to increase the mean subsidence rate over the desert. This in turn increases the aridity and enhances deflation of more dust from the desert surface. Since a thick grass cover develops naturally inside animal exclosures on the Rajasthan Desert, it is evident that much of the desert itself may be man-made. Indeed, there is some evidence that it was made by the Indus civilization [14]. Thus we must consider the possibility that the 'natural' dust load of the atmosphere over the Indian desert is largely man-made, or at least the result of animal domestication.

Vukovich and Chow [15] have shown in their numerical simulation of the atmosphere that a distributed heat source can stabilize the long waves in the upper air-streams. The results of the Indian dust study suggest that the effect of the dust-enhanced diabatic cooling is of such a magnitude that it should be taken into account in such calculations. Yet *there are no systematic observations of dust densities and distribution, and apparently no plans for such observations* contained in the Global Atmospheric Research Program or the World Weather Watch!

There have been many papers written about the climatic effects of sunspots and the rising carbon dioxide content of the atmosphere, but very few on the global climatic effect of dust other than volcanic.

The theory is not complete but there are enough data to at least get a statistical hint of the order of magnitude of the effects. The well-known equation for the radiative equilibrium of the earth is

$$S\pi R^2(1-\alpha)=4\pi R^2 I_t$$

where S = intensity of solar beam at top of the atmosphere, i.e. the 'solar constant'; R = radius of the earth; α = global albedo; I_t = outward radiation intensity at top of the atmosphere.

The usual values of S and α agree quite well with satellite measurements of I_t [16]. We can approximate the surface temperature through

$$I_t=I_0-\Delta I=\varepsilon\sigma T_0^4-\Delta I$$

where I_0 = the emitted radiation of the earth's surface; T_0 = 'mean' surface temperature of the earth; $\varepsilon\sigma$ = the effective emissivity of the surface times the Stefan-Boltzmann constant and ΔI = the difference between the upward radiation from the surface of the earth and the outward radiation at the top of the atmosphere, i.e. the 'greenhouse effect'.

It is assumed that S is somewhat dependent on the sunspot number N, α is in part a function of the turbidity τ [17], and ΔI is in part dependent on the carbon dioxide content of the atmosphere C. Then the temperature is a function of N, τ, and C and the variance due to each may be assessed. This was done in part by Mitchell [18] for carbon dioxide and sunspots. Taking Davitaia's dust-fall measurements [3] as an approximation of the amount of smaller particles resident in the atmosphere, we find that over the past eight decades the three independent variables, carbon dioxide, dust, and sunspots contributed 71%, 8% and 5% respectively to the variance of T_0^4. Examination of Figure 6 shows that until 1930 there was little variance of the dust, but since that time the increase of dust has dominated and more than negated the increase of carbon dioxide, terminating the warming and producing a decline of world temper-

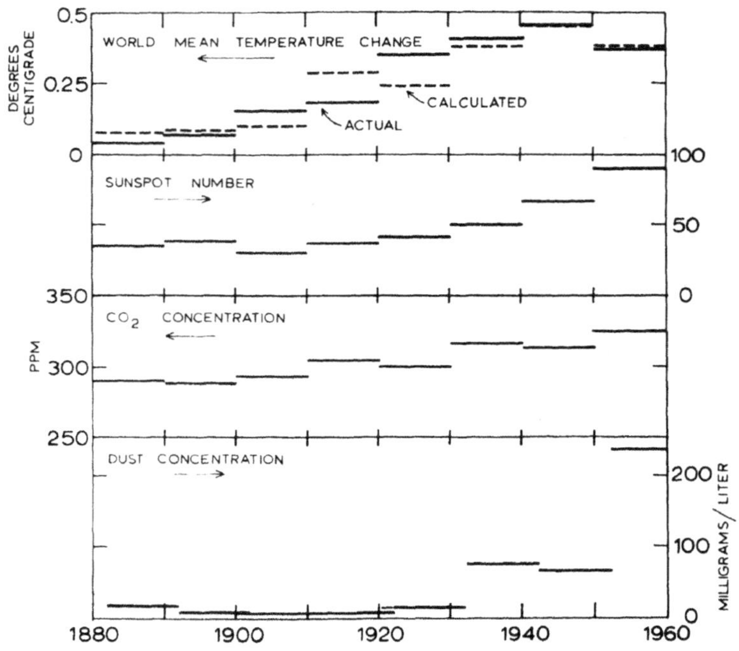

$$\Delta T\,(^\circ C) = -3.546 + 0.012\,CO_2 - 0.002\ \text{Dust} + 0.006\ \text{Sunspots}$$

Fig. 6. Trend of mean world temperature change (18), CO_2 concentration (18), sunspot number (18) and dust in the Caucasus (3). Stepwise multiple regression of above variables, with temperature change as dependent variable.

ature. It has been argued that the carbon dioxide increase was the work of man also, in the burning of fossil fuels, but as Deevy [19] has pointed out, the radiocarbon evidence suggests that the increase does not appear to be of fossil carbon, thus is more likely to be due to increased oxidation of plant materials such as soil humus and bogs. Yet even here one must recognize the role of increased cultivation, forest clearing, and bog draining. In either case man is inadvertently modifying the climate.

There is another effect of increased turbidity that is more important locally than the mean world temperature change, and that is the effect on the meridional radiation gradient and thus on the thermal Rossby number and the circulation pattern. Increased turbidity should reduce the meridional radiation gradient and thus weaken the westerlies. Lamb [20] has shown that such weakening has been characteristic of the 1960's. This change of gradient can have locally profound climatic effects.

In the preceding paragraphs we have not considered, for lack of enough information, other possible widescale climatic modifications due to human activity – such as worldwide cloud seeding by automobile exhaust [8], the effect of jet contrails on cloud cover, or the effect of atmospheric turbidity in reducing the ultraviolet intensity in the Vitamin D band as hinted at by the work of Sastri and Das [21].

It is possible, however, to make a crude estimate of the effect of contrails. Taking

3000 as the number of jet aircraft in the air, averaging 500 mi/hr, 50% making contrails, which last an average of 2 hours and spread to a width of ½ mile we have

$$3000 \times 500 \times 0.5 \times 2 \times 0.5 \text{ mi}^2 \text{ of contrails.}$$

Dividing by the area of the region in which most of these aircraft are operating we find a 5–10% increase in cirrus in the North American-Atlantic-Europe area or about a twentieth of this for the world. This is not negligible!

5. The Future

It appears that population growth, mechanization, and industrialization have now made man the equivalent of other natural processes in his effect on climate. The industrial revolution is still underway in large parts of the world, and if we can attribute either the carbon dioxide increase and/or the recent increase of atmospheric turbidity to human activities it appears that there is little that any one nation can do to reverse the trend. Nevertheless, if the analysis of this paper is correct or even nearly correct, it behooves us to study the problem much more intensively than we have. We would be pleased to be proven wrong. It is too important a problem to entrust to a half-dozen part-time investigators.

Even in modern industrial cultures, man is too vulnerable to environmental conditions to ignore even what may appear to be trivial problems. For example, if there are 300 supersonic transports normally in the air travelling 1500 mi/hr they will make 4.5×10^5 XYZ mi^2 of cloud cover, where X is the percentage making contrails, Y is the average duration of the contrails in hours, and Z is the average width of the contrails in miles. It has been stated that at design altitude the value of X is zero. Air Force experience suggests that this is not true. If for some reason these aircraft do not operate at design altitude, but near the tropopause instead, the value of X may approach 1.0, and if Y is 25 hrs, and Z one mile, the resulting cloud cover over the region of operation of most of the SST's would approach 100%. We would like our grandchildren to experience blue skies more often than on rare occasions!

References

[1] Peterson, J. T. and Bryson, R. A.: 1968, *Proc. of First Nat'l Conf. Wea. Mod.*, Albany, N.Y.
[2] Risebrough, R. W., Huggett, R. J., Griffen, J. J., and Goldberg, E. D.: 1968, *Science* **159**, 3820.
[3] Davitaia, F. F.: 1965, *Trans. Soviet Acad. Sci., Geogr. Ser.* No. 2.
[4] Davitaia, F. F.: 1968, Personal Communication, Georgian Acad. of Science, Tbilisi, U.S.S.R. He also reported a similar trend to dust concentration from the Altai Mts.
[5] McCormick, R. A. and Ludwig, J. H.: 1967, *Science* **156**, 1358.
[6] Peterson, J. T. and Bryson, R. A.: 1968, *Science* **162**, 3849.
[7] Changnon, Jr., S. A.: 1968, *Bull. Amer. Met. Soc.* **49**, 1
[8] Schaefer, V. J.: 1968, *Proc. of First Nat'l Conf. Wea. Mod.*, Albany, N.Y.
[9] Lettau, H. and Lettau, K.: 1969, *Tellus* **21**, 208.
[10] Landsberg, H. E.: 1962, *Symposium – Air Over Cities*, Sanitary Eng. Center Tech. Rep. A62-5, Cincinnati, Ohio.
[11] Bryson, R. A. and Baerreis, D. A.: 1967, *Bull. Amer. Met. Soc.* **48**, 3.

[12] Das, P. K.: 1962, *Tellus* **14**, 2.
[13] Nagatani, R. M.: 1968, M.S. thesis, Dept. of Meteor., Univ. of Wis., Madison, unpublished.
[14] Singh, G.: 1968, personal communication, Birbal Sahni Institute of Paleobotany, Lucknow, India.
[15] Vukovich, F. M. and Chow, C. F.: 1968, Research Triangle Inst., Research Triangle Park, N.C., Final Rep. Cont. AF19(628)-5834.
[16] Vonder Haar, T.: 1968, Ph.D. thesis, Meteor. Dept., Univ. of Wis., Microfilm Library, Ann Arbor, Mich.
[17] Ångstrom, A.: 1962, *Tellus* **14**, 435.
[18] Mitchell, Jr., J. M.: 1961, *Ann. N.Y. Acad. Sci.* **95**, 1.
[19] Deevey, Jr., E. S.: 1958, *Sci. Amer.* **209**, 10.
[20] Lamb, H. H.: 1966, *Geogr. J.* **132**, 2.
[21] Sastri, V. D. P. and Das, S. R.: 1968, *J. Opt. Soc. Amer.* **58**, 3.
[22] Powell, R. P.: 1968, *Sci. Amer.* **219**, 6.

For Further Reading

1. 'All Other Factors Being Constant...', *Weatherwise* **21**, No. 2 (April, 1968), 56–61.
2. L. J. Battan, *The Unclean Sky*, Doubleday & Company, New York, 1966.
3. R. A. Bryson and J. E. Kutzbach, Air Pollution, Assn. Amer. Geogr. Resource Paper No. 2, Washington, D.C., 1968.
4. R. G. Ridker, *Economic Costs of Air Pollution*, Frederick A. Praeger, New York, 1967.
5. A. C. Stern (ed.), *Air Pollution*, Vols. 1 and 2, Academic Press, New York, 1962.

A PRELIMINARY EVALUATION OF ATMOSPHERIC POLLUTION AS A CAUSE OF THE GLOBAL TEMPERATURE FLUCTUATION OF THE PAST CENTURY

J. MURRAY MITCHELL, JR.*

Department of Atmospheric Sciences, University of Washington, Seattle, Washington, U.S.A.

Abstract. Two globally extensive forms of atmospheric pollution (carbon dioxide and particulate loading) are each considered from the viewpoint of long-term changes in their world-average abundance, and the relevance of these to the observed fluctuation of planetary average temperatures in the past century.

On the basis of significant new data on atmospheric CO_2 that has become available in recent years, it is reliably estimated that about half of all fossil CO_2 released by man since the 19th century has remained in the atmosphere, and that the doubling time of atmospheric CO_2 accumulation from this source is about 23 years. The present-day CO_2 excess (referred to 1850) is estimated at 11 %; the excess is projected to increase to 15 % by 1980, 20 % by 1990, and 27 % by 2000 A.D. Changes of mean atmospheric temperature due to CO_2, calculated by Manabe *et al.* as 0.3°C per 10 % change in CO_2, are sufficient to account for only about one third of the observed 0.6°C warming of the earth between 1880 and 1940, but will probably have become a dominant influence on the course of planetary average temperature changes by the end of this century.

The secular increase of global atmospheric particulate loading by human activity is estimated, and compared with a construction of the secular variability of stratospheric dust loading derived from data on volcanic activity since 1850. It is concluded that the total human-derived particulate load is at present comparable to the *average* stratospheric dust load from volcanic eruptions, but that the variations of human-derived loading are an order of magnitude less than those of volcanic dust loading. For reasonable estimates of the thermal cooling effect of dust load increases, it is inferred that secular cooling due to human-derived particulate loading is currently of the order of 0.05°C per decade. Although changes of total atmospheric dust loading may possibly be sufficient to account for the observed 0.3°C-cooling of the earth since 1940, the human-derived contribution to these loading changes is inferred to have played a very minor role in the temperature decline.

Of the two forms of pollution, it appears that the carbon dioxide increase is several times more influential in raising planetary temperatures than the human-derived particulate loading increase is in lowering planetary temperatures. If, however, the doubling times of particulate loading and CO_2 accumulation remain unchanged in the future (15 to 20 years and 23 years, respectively), a further warming of the order of 1°C or more will culminate some time after 2000 A.D., followed by a net cooling as the particulate loading effect ultimately overtakes the CO_2 effect. Other environmental agencies, presumably natural ones, are required to account for the main part of the observed fluctuation of world-average temperatures during the past century, and will continue to exert an important nfluence on climate in the future.

1. Introduction

In recent years there has been growing speculation that air pollution is responsible for world-wide disturbances of weather and climate. There are two basic lines of reasoning that have fed such speculation.

One line of reasoning is of a circumstantial nature. It begins with the realization that, according to various meteorological evidence, the large-scale climate of the earth

* Permanent affiliation: Environmental Science Services Administration, U.S. Department of Commerce, Silver Spring, Md., U.S.A.

has fluctuated during the past century. This fluctuation happens to have occurred at a time when modern industrial man was first emerging as a powerful manipulator of his physical environment. Inasmuch as the fundamental cause or causes of this climatic fluctuation are not yet identified, it is argued that – unless or until demonstrated otherwise – one or more inadvertent byproducts of man's (historically unprecedented) activities cannot be ruled out as possible contributory causes.

The other line of reasoning is more direct, and in some respects more compelling. This one takes off from the fact that locally heavy concentrations of air pollution, such as we find in cities and heavy-industry areas, have undeniable effects on local weather and climate. Among these effects are reduced solar radiation and visibility, increased fog and low cloudiness, and perhaps anomalous temperature conditions as well [1]. Such effects, incidentally, are consistent with theoretical principles of atmospheric physics, and are predictable to some degree. Now since certain types of air pollutants have atmospheric residence times measured in weeks or more [2], these pollutants are obviously capable of being spread by atmospheric circulations, in diluted concentrations, over very wide geographical areas. It presumably follows, then, that certain meteorological effects of these pollutants should extend over equally wide geographical areas, and, in the case of at least one pollutant (carbon dioxide, whose atmospheric residence time is measured in years rather than months), the effects should be strictly global. From a qualitative viewpoint, such reasoning is obviously sound. It is left for us first to establish the prevailing concentration of each geographically extensive form of air pollution, together with its changes over the years, and second to determine as reliably as possible the quantitative meteorological or climatological effects of each such pollution form.

Before we comment further on the idea that man is responsible in any way for altering world climate, it would seem prudent for us to keep in mind that world climate has unquestionably varied throughout the earth's history, and sometimes dramatically so. It has, in fact, varied significantly during earlier centuries of the Christian Era [3], long before man's capacity for influencing his physical environment on a large scale had developed. The obvious implication of this is that world climate is clearly capable of variation through *natural* causative agencies, whatever those agencies may specifically be. Thus, the climatic fluctuation of the past century could also be attributable entirely to natural causes, as for example some type of non-linear thermodynamic interaction between the atmosphere and the oceans, or a more remote environmental disturbance that can be expected to influence climate (for example variable radiation from the sun, or episodes of unusually great volcanic activity) [4].

On the other hand, if man is not yet in nature's league as a potent climate-regulating force, he is almost certainly destined to become such a force in the rather near future. Of special concern in this connection is the dangerous circumstance that man may well arrive at that point *inadvertently* before he arrives there deliberately, and that he will find himself unequipped to arrest or reverse undesirable climatic developments that he may have set in motion unwittingly. From any point of view, therefore, it is a matter of some urgency that we identify the causes of modern-day climatic instability, and make an accurate determination of man's impact on world climate both present and future.

2. Scope of This Discussion

There are many different aspects of the problem of large-scale air pollution/climate relationships that are in need of much more intensive investigation. Some involve the possible influence of pollutants on cloudiness and precipitation, along such lines as those Dr. Schaefer has considered in this symposium. Others involve the planetary heat budget and the equilibrium temperature of the earth's surface, such as Dr. Bryson and Dr. Manabe have considered.

My remarks will be confined to two specific forms of pollution, and their apparently competing impacts on the thermal climate of the earth. These are (1) carbon dioxide and (2) atmospheric particulates (dust and smoke). Both have already been introduced by other contributors to this symposium, so in my remarks I will obviously be treading over some of the same ground. I do so, without apology, for two reasons. First, it has recently become possible to pin down the carbon dioxide factor with considerably more reliability than ever before, and so a fresh look at the CO_2 situation seems particularly appropriate. Second, with regard to the atmospheric dust loading factor, I intend to take this up in some detail from a rather different perspective than Dr. Bryson's. This perspective is one in which we recognize certain *naturally* induced changes of atmospheric particulate loading simultaneously in progress, and take advantage of what can be surmised about the thermal effect of these natural changes to estimate the analogous thermal effect of man-made loading changes.

Let us first turn briefly to the evidence for changes of the average temperature of the earth, which are to be interpreted as one (and, practically, very important) manifestation of the variations of the planetary heat budget during the past century.

3. Recent Planetary Temperature Trends and Their Significance

By analysis of climatological data for stations distributed as uniformly as possible over the earth's surface, it can be established that the mean temperature of the whole planetary atmosphere, at least in its surface layers, has fluctuated systematically during the past century [5]. The data reflect a net world-wide warming of about 0.6 °C (1.0 °F) between the 1880s and the 1940s, followed by a net cooling of about 0.2 °C (0.3 °F) between the 1940s and 1959, the most recent year of data in the original analysis. Additional data for 1965–67, as published by Scherhag [6], were later incorporated into the analysis in order to estimate the further movement of world mean temperature after 1960 (see Figure 1). On this tentative basis, it appears that the cooling trend which first set in during the 1940s has continued essentially up to the present time, and that the net temperature drop in the last quarter-century has now accumulated to about 0.3 °C (0.5 °F). To date, then, about half of the warming that occurred during earlier decades of the century has been erased by subsequent cooling (Figure 1). One cannot say offhand whether or how long this cooling will continue in the future.

The observed fluctuation in world mean temperature is believed to indicate a systematic change in the planetary heat budget during the past century. Although the magnitude of the fluctuation may appear extremely small, in comparison to the familiar seasonal temperature changes for example, it is symptomatic of larger

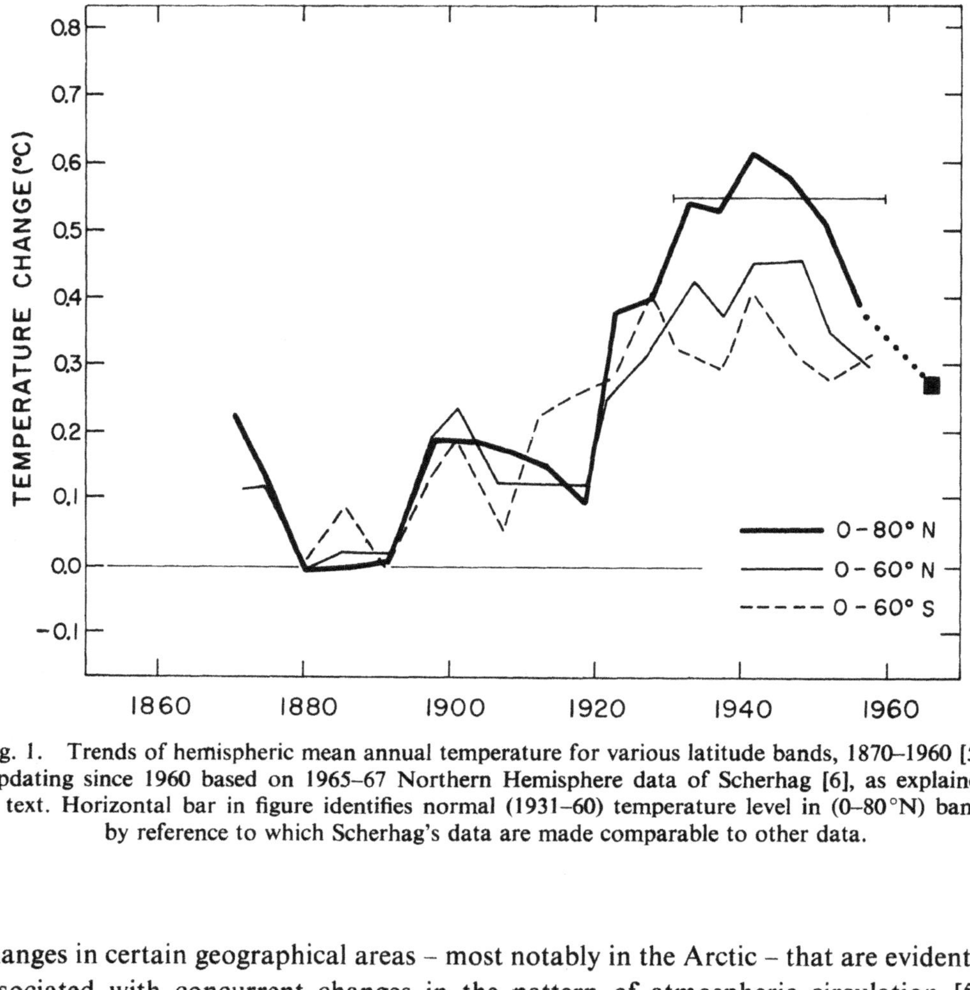

Fig. 1. Trends of hemispheric mean annual temperature for various latitude bands, 1870–1960 [5]. Updating since 1960 based on 1965–67 Northern Hemisphere data of Scherhag [6], as explained in text. Horizontal bar in figure identifies normal (1931–60) temperature level in (0–80°N) band, by reference to which Scherhag's data are made comparable to other data.

changes in certain geographical areas – most notably in the Arctic – that are evidently associated with concurrent changes in the pattern of atmospheric circulation [5]. Moreover, the magnitude of the fluctuation, of the order of 0.6 °C, is an appreciable fraction of the planetary temperature difference of about 6 °C that is believed to have distinguished glacial from interglacial conditions during the Pleistocene Ice Age [7]. Indeed, the temperature changes of the past century have been paralleled by well-documented and widespread changes in alpine glacier movement, Arctic pack-ice cover, desert margins, world sea levels, and floral and faunal limits. In such respects, the temperature fluctuation reflects a climatic disturbance of no inconsiderable practical significance. Our ignorance of the future course of that disturbance is a strong motivation for us to establish its causes.

4. The Secular Increase of Atmospheric Carbon Dioxide

A. DISCUSSION OF THE EVIDENCE FOR INCREASING CO_2

Since late in the 19th century, man has released carbon dioxide to the atmosphere at an ever-accelerating rate through the combustion of fossil fuels (primarily lignite, coal, petroleum and natural gas). The present rate of CO_2 production by man is estimated at 9×10^9 metric tons per year [8]. Until recently, however, it was not clear how much of this gas has accumulated in the atmosphere, and how much has been removed by a net transfer to the oceans and to the terrestrial biomass.

Various efforts have been made by Callendar and others [9] to estimate the actual CO_2 accumulation in the atmosphere by means of a statistical comparison of historical measurements of atmospheric concentrations available during the past 100 years. These have suggested a systematic increase from a representative 19th century base level of about 290 parts per million (ppm) to about 330 ppm by 1960, an increase of 14% in the period. If these figures are correct, they imply that nearly 75% of all CO_2 released by man has remained in the atmosphere. But according to Keeling, a proper consideration of the effect of differences of instrumentation reveals that these figures are too large [10].

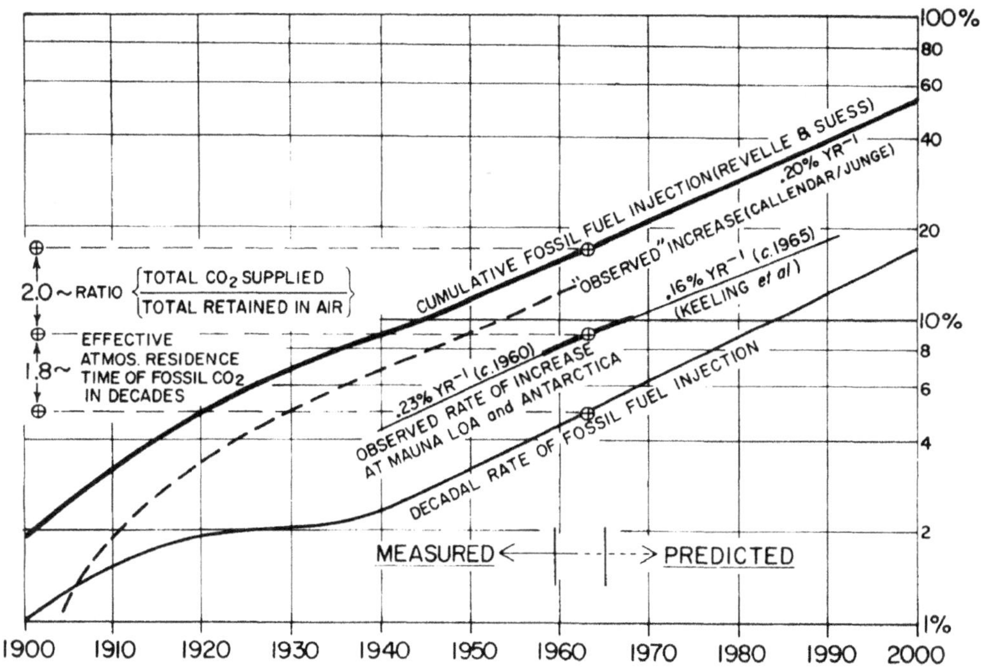

Fig. 2. Secular increase of atmospheric carbon dioxide due to fossil fuel combustion. Top curve is increase assuming all fossil CO_2 has remained in atmosphere [12]. Dashed curve is illustrative of earlier estimates of actual CO_2 increase in atmosphere. Trend segments in middle of figure identify mean level and rates of increase of CO_2 considered representative of true world wide conditions in period 1958–68 [10, 11]. Bottom curve is decadal injection rate of fossil CO_2 [12]. Best estimate of actual increase is between 40 and 50 % of that indicated by top curve.

Keeling *et al.* [11] have substantially refined such estimates on the basis of a CO_2 monitoring program begun in 1958 in cooperation with the U.S. Weather Bureau and ESSA. In this program, precision gas-analyzer measurements have been made at Mauna Loa Observatory in Hawaii and at the Pole Station in Antarctica. Inasmuch as the measuring sites are at high altitudes and are well removed from local upwind sources of contamination, CO_2 concentrations observed there rarely deviate by more than about 1 ppm, on an average, from their true ambient values. The latter, in turn, are expected to remain within 2 or 3 ppm of the worldwide annual average atmospheric concentration, owing to the rapid mixing of the gas by planetary-scale air motions.

It is therefore highly significant that Keeling and his colleagues have detected a systematic increase in CO_2 concentration since 1958, at the rate of about 0.7 ppm per year before 1965 and at the slightly slower rate of about 0.5 ppm per year since 1965 [10, 11]. (Keeling believes that the recent slowing of the trend rate is a transient effect having no long-term significance.) The average of these rates, about 0.6 ppm yr^{-1} (0.20% yr^{-1}), coincides almost exactly with the contemporary rate of relative increase of total fossil CO_2 injection into the atmosphere as estimated by Revelle and Suess [12]. (See Figure 2.) Moreover, the annual average CO_2 levels measured concurrently at Mauna Loa and the Pole Station agree within 1 ppm (314.0 ppm and 313.4 ppm, respectively, in 1960). Taken together, these statistics lead us with some confidence to the following conclusions.

(1) The rate of the secular increase of atmospheric CO_2 can now be reliably established at about 0.20% per year, corresponding to a doubling time of about 23 years in the excess above 19th century levels.

(2) The average planetary atmospheric CO_2 concentration circum 1960 was very close to 315 ppm, or 8% above the generally accepted 19th-century base concentration of about 290 ppm. This is a substantially smaller increase than Callendar's earlier estimates [9], but at the same time a substantially larger increase than that inferred from the rate of radiocarbon dilution measured in tree wood (the Suess effect) [13].

(3) The observed secular increase of CO_2 is almost certainly attributable to fossil fuel combustion, and of all the fossil CO_2 that has been injected into the atmosphere from this source since the 19th century, it is now rather certain that between 40 and 50% has remained in the atmosphere.

(4) Consistent with the foregoing conclusions, it is estimated that relative to a 19th century base level of 290 ppm the atmospheric CO_2 content had increased 5% by 1944, and 10% by 1967. Similarly, the increase is projected to accumulate to 11% by 1970, 15% by 1980, 20% by 1990, and 27% by 2000 A.D.

B. SIGNIFICANCE TO THE THERMAL STATE OF THE ATMOSPHERE

Undoubtedly the most reliable determination yet available of the effect of changes of atmospheric CO_2 on the equilibrium temperature distribution in the atmosphere is that based on the numerical investigations of Manabe *et al.* [14]. According to these authors, increases of CO_2 result in a warming of the entire lower atmosphere, the amount of warming being dependent in part on whether the atmosphere is held at a fixed absolute humidity or at a fixed relative humidity during the CO_2 change. For the assumption of fixed *absolute* humidity, together with conditions of surface albedo, cloudiness, solar radiation, and other parameters chosen as typical of the mid-latitudes in an equinoxial season, the temperature effect is such that a 10% increase of CO_2 concentration (from 300 to 330 ppm) would lead to a warming of about 0.2 °C (0.3 °F). For the assumption of fixed *relative* humidity, all other conditions remaining the same, the temperature effect of a 10% CO_2 increase is nearly doubled to 0.3 °C (0.5 °F). Manabe and Wetherald [14] reason that the latter assumptions are probably more realistic, in which case it is possible to ascribe about one third of the observed world-wide warming trend between 1880 and 1940 to the secular increase of CO_2. We are apparently required to look elsewhere to account for the remaining two thirds of this warming. On the other hand, since the theoretical relationship between CO_2 concentration and equilibrium temperature is nearly linear in the limited range of CO_2 variations involved, we should note that the rate of warming due to CO_2 was probably greater since 1940 than it was at any earlier time. In fact, with a CO_2 doubling time of 23 years which we had estimated previously, somewhat more than half of the net warming to date that is attributable to CO_2 since the 19th century (0.3 °C in 1969) is expected to have occurred since 1940. This, of course, was during the time when world-mean temperatures are estimated to have *declined* by about 0.3 °C. Thus it appears that the climatic cooling mechanism coming into operation during the past quarter century has been roughly three times more influential on temperature than the CO_2 effect in the same period.

In summary, I suggest that, according to the best information now available, probably not more than one third of the planetary temperature disturbance of the past century is attributable to variations of atmospheric CO_2. Other mechanisms are evidently required to account for part of the warming observed between 1880 and 1940, as well as for the cooling observed since 1940 which has occurred in spite of further warming contributions by CO_2 in that period. The temperature contribution of CO_2 changes anticipated in the future, neglecting all other mechanisms of climatic change, will consist of a further warming (above 1969 temperature levels) of about 0.1 °C (0.2 °F) by 1980, 0.3 °C (0.5 °F) by 1990, and 0.5 °C (0.8 °F) by 2000 A.D. It is therefore likely that, if other causative factors in climatic change do not also become more important in the future than in the past, carbon dioxide will win out by the end of this century as the dominant factor in determining the future course of planetary temperature.

5. Changes of Atmospheric Dust Loading

A. COMPARISON OF HUMAN AND NATURAL CONTRIBUTIONS TO ATMOSPHERIC PARTICULATE LOAD

Dr. Bryson has cited evidence that a dramatic increase in particulate loading of the atmosphere has occurred in the last few years. He has suggested that this increase is due to human activity, and that it is the probable cause of the latter-day decline in planetary average temperature already noted. In view of the relatively rapid rate at which this cooling has proceeded since 1940, it behooves us to check more fully into Bryson's hypothesis with all its ominous implications for the future. In the following discussion, I shall not be arriving at any firm conclusions as to the validity of the hypothesis. However, there are two important considerations that bear on the matter, which have not been sufficiently aired. One consideration is the extent to which the atmospheric particulate loading may have varied in recent years through agencies beyond man's control. The other consideration, in the face of various difficulties that preclude an accurate theoretical assessment at this time of the thermal effects of variable atmospheric particulate loading, is concerned with the possibility of estimating the magnitude of these effects by empirical inference.

In this symposium, Goldberg has estimated the present rate of injection of smoke particles into the world atmosphere by human agencies at 2×10^7 metric tons per year [8]. Virtually all of this material is injected into the tropospheric layer of the atmosphere. The tropospheric residence time of most chemically inert particulates is of the order of 0.1 year [2]. Adopting this figure for the average residence time of the smoke particles, we can arrive at an estimated average total planetary atmospheric smoke loading that corresponds to Goldberg's value of the injection rate. This turns out to be 2×10^6 metric tons. In the absence of any direct measure of the rate at which this total loading has increased during the past century, it would seem reasonable to suppose that it has increased in direct proportion to the injection rate of fossil CO_2 estimated by Revelle and Suess [12]. On this basis, the secular increase of particulate loading due to human activity, in metric tons, is as shown in Figure 3.

In addition to human particulate loading, there is always, of course, a substantial atmospheric loading by dust, smoke, and other particles that are derived from natural

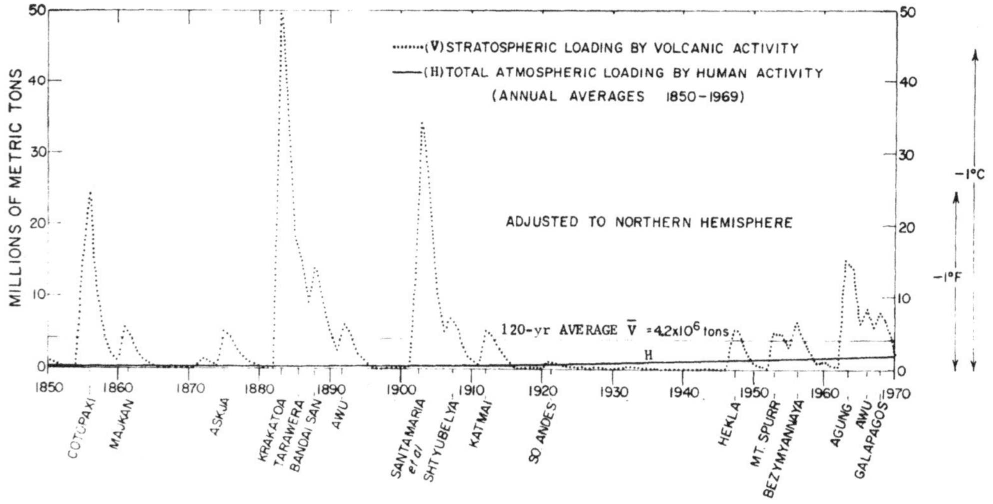

Fig. 3. Estimated chronology of world wide atmospheric particulate load by volcanic activity (stratospheric loading only, dotted curve) and by human activity (heavy solid curve). The 120-year average of loading by stratospheric volcanic dust is added for comparison (thin solid line). Estimated calibration of figure in terms of planetary temperature influence is shown outside margin at right.

TABLE I

Volcanic eruptions since 1855

Date	Name and location			Severity class*
1855	Cotopaxi, Ecuador	1°S	78°W	1–1/2
1856	Awu (Awoe)	3.5°N	125.5°E	2
1861	Makjan, Molucca Is.	.5°N	127.5°E	2
1870	Ceboruco, Mexico	21°N	105°W	3
1872	Vesuvius	41°N	14°E	3
1872	Merapi, Java	7.5°S	110°E	3
1875	Askja (Vatna Jökull) Iceland	65°N	17°W	2
1877	Cotopaxi, Ecuador	1°S	78°W	3
1883	Krakatoa	6°S	105.5°E	1
1883	St. Augustine, Alaska	59.5°N	153.5°W	3
1883	Bogoslov, Aleutians	54°N	168°W	3
1885	Falcon Island	20°S	175°W	3
1886	Tarawera, N.Z.	38.5°S	176.5°E	2
1886	Niafu, Tonga Is.	16°S	175.5°W	3
1888	Bandai San, Japan	38°N	140°E	2
1888	Ritter Is.	5.5°S	148°E	2
1890	Bogoslov, Aleutians	54°N	168°W	3
1892	Awu (Awoe)	3.5°N	125.5°E	2
1902	Mont Pelée, Martinique	15°N	61°W	2
1902	Soufrière, St. Vincent	13.5°N	61°W	2
1902–04	Santa Maria, Guatemala	14.5°N	92°W	1–1/3
1907	Shtyubelya, Kamchatka	52°N	157.7°E	2
1911	Taal, Luzon	14°N	121°E	3
1912	Katmai, Alaska	58°N	155°W	2
1913	Colima, Mexico	19.5°N	104°W	3
1914	Sakurashima, Japan	31.5°N	131°E	3
1921	Andes (Chile-Arg. Border)	≃ 30°S	≃ 70°W	3
1929	Asama, Japan	36.5°N	138.5°E	4
1931	Kluchev, Kamchatka	56°N	160.5°E	4
1932	Quizapu, Chile	35.5°S	70.5°W	3
1947	Hekla, Iceland	64°N	19.5°W	2
1953	Mt. Spurr, Alaska	61°N	153°W	2
1955	Ranco Puyehue, Chile	40°S	72°W	3
1956	Bezymyannaya, Kamchatka	56°N	160.5°E	2
1960	Puntiagudo *et al.*, So. Chile	39–45°S	72°W	3
1963	Gunung Agung, Bali	8.5°S	115.5°E	1–1/2
1963–65	Surtsey, Iceland	63°N	20.5°W	3
1966	Awu (Awoe)	3.5°N	125.5°E	2
1968	Fernandina I., Galapagos	0.5°S	92°W	2

* Severity class reflects the estimated order of magnitude of the total mass of material ejected.

Class 1: 10–1 $km^3 \sim 10^{10}$ metric tons
Class 2: 1–0.1 $km^3 \sim 10^9$ metric tons
Class 3: 0.1–0.01 $km^3 \sim 10^8$ metric tons
Class 4: 0.01–0.001 $km^3 \sim 10^7$ metric tons

Fractional classes represent compromises believed to be appropriate. Data based primarily on unpublished table of H. H. Lamb.

sources [2]. Many of these particles are in the same size range (0.1 to 1μ) as the smoke introduced through human agencies. This is well recognized as the size range having the greatest influence on the absorption and scattering of solar radiation in the atmosphere, and it is consequently of special importance to the planetary heat budget.

Here we are interested not so much in the total particulate loading of the atmosphere as we are in the *variable* part of the total loading. Although all sources of atmospheric dust are likely to be variable, there is one source in particular that is extremely variable. That source is volcanic dust, of the fine-grain form that is occasionally injected in large quantities into the upper levels of the atmosphere by violent paroxysmal eruptions, where it is subsequently spread over large areas of the world by atmospheric winds before it can settle back to earth.

Based mainly on an unpublished list of volcanic eruptions during the past five centuries, prepared by H. H. Lamb of the British Meteorological Office [15], I have taken a stab at reconstructing the chronology of total atmospheric loading by volcanic dust during the past century. Admittedly, this is a rather hazardous undertaking, since there were undoubtedly some major eruptions during the period that were unobserved (and therefore not listed), and since there are very great uncertainties involved in evaluating eruptions according to the mass of fine dust explosively ejected.

The reconstruction of the atmospheric dust load chronology was based on the eruption dates listed in Table I, and on the order-of-magnitude estimates of dust ejection in each eruption also shown in the table. The chronology was then developed according to the following assumptions: (1) The portion of the total ejected mass that reached the stratosphere as fine dust was assumed to be 1% for all eruptions. (2) Only the mass of dust assumed to reach the stratosphere was considered in developing the chronology. The remaining 99% of the ejected mass was neglected, on the basis that is consisted mainly of large blocks or particles with rapid fall speeds, and that only a minute part of it could remain suspended long enough in the lower atmosphere to be dispersed over areas comparable to the earth's total surface area. (3) A residence time of 14 months [16] was assigned to all dust assumed to reach the stratosphere, regardless of geographical location. (4) The chronology was adjusted to the Northern Hemisphere by assigning all eruptions south of 16°S latitude an order of magnitude less severity than shown in Table I. (5) The dust load was derived as an average load in each calendar year, the contribution of a new eruption being taken as one half of its estimated stratospheric injection mass in the first year, and as the total stratospheric injection mass attenuated by the 14-month residence time thereafter. This procedure was intended to reduce the first year's loading by new eruptions, in allowance for the time lags involved in the dispersal of stratospheric dust over very large downwind distances and in waiting for seasonal changes of stratospheric circulation that promote rapid dust dispersion into other latitudes. The resulting chronology is shown in Figure 3.

If the above assumptions are even approximately valid, we may conclude from Figure 3 that the total particulate loading of the atmosphere by human activity is beginning to approach the long-term average stratospheric dust loading by volcanic activity. It follows that certain climatic effects of volcanic dust in the atmosphere – whatever these may specifically be – are likewise being approached by comparable effects of smoke and dust derived from human activity. On the face of it, this conclusion seems reasonable.

However, it is obviously the *changes* of dust loading over the years that are relevant to comparatively abrupt world-wide climatic changes, such as the cooling after 1940. According to Figure 3 the volcanic load has been increasing much more rapidly in the past quarter-century than the human-derived particulate loading. The series of latter-day eruptions, beginning with Hekla in 1947, marked the introduction of a large dust load that coincided closely with the period of abruptly reversed global temperature change. In order to find earlier episodes of similarly large volcanic dust loading, one has to look back as far as 1915, when, incidentally, world mean temperatures are known also to have been relatively low.

It therefore seems that if the two kinds of anomalous particulate loading of the atmosphere have comparable impacts on the planetary heat budget (as theoretical considerations would indicate they should), then nature rather than man is to be held primarily responsible for such relatively abrupt cooling phases in world climate as that which has developed in recent years.

With this background, let us now take a closer look at other evidence cited by Dr. Bryson that there have been rapid increases of total atmospheric turbidity in recent years.

First there is the evidence based on dust fall in the high Caucasus as originally reported by Davitaia [17], which Bryson and Wendland have compared with a measure of Soviet industrial expansion in their Figure 2 (this volume, p. 132). The general correspondence between the two curves in that figure is highly suggestive of a real relationship which I would not question. On the other hand, the extremely rapid rate of industrial growth shown there is greatly in excess of the growth rate considered representative of the world as a whole (as reflected, for example, in the world-wide CO_2 production estimates of Revelle and Suess). This suggests to me that much of the dust fall weighed by Davitaia probably consisted of relatively large particles that were scavenged from the atmosphere in orographic precipitation over the Caucasus, and that were quite local in origin [2]. If this is the case, such data could hardly be interpreted as a reliable *world-wide* index of atmospheric dust loading, such as we require to compare with the world-wide temperature data.

Next there is the evidence based on measurements of atmospheric turbidity at the Mauna Loa Observatory in Hawaii, shown by Bryson and Wendland in their Figure 4 (this volume, p. 133) and published earlier by Peterson and Bryson [18]. It is unfortunate that the authors have not discussed these measurements more fully, and explained precisely what they mean when they interpret such measurements as showing a 30% increase of turbidity in the period 1957–67. Their basic data are expressed in terms of the change of atmospheric extinction of direct solar radiation at the altitude of the Observatory (3500 m), in units of the total extinction by a clear atmosphere above the same altitude. If these data are referred to sealevel, with allowance for additional extinction due to typical dust loading below the altitude of the Observatory (equivalent to 1.0 to 1.5 clear atmospheres), then the measured turbidity increase of 30% is found to be equivalent only to a 10 to 15% increase in the total dust load above sealevel. This equivalence is, of course, valid only if there were no change in dust load below the observatory level, which is a reasonable assumption if the measured turbidity change is attributable primarily to a stratospheric aerosol such as volcanic dust.

That the arrival of a volcanic dust pall from the Agung, Bali, eruption of 1963 was the probable cause of the very rapid turbidity increase in that year at Mauna Loa is acknowledged by Bryson and Wendland. However, the fact that the turbidity remained very high for a number of years following the Agung eruption seemed inconsistent with the normal stratospheric residence time for volcanic dust, which is generally 14 months for tropical eruptions [16]. This led Bryson *et al.* to suppose that the high turbidity levels after 1964 were due to sources other than volcanic dust, and represented the continuation of a background increase that seemed to be in evidence also before 1963. Peterson and Bryson [18] went on to propose that this background increase may have been attributable to human activities.

There are two circumstances, however, that militate against such an interpretation of the Mauna Loa turbidity trend. First, the renewed high turbidity levels in 1966 and 1967 were likely to have been associated with a second volcanic eruption of major intensity, to wit, the eruption of Awu on Great Sangihe Island (3.5°N, 125.5°E) in August of 1966 [15]. Second, if the net trend at Mauna Loa were attributable to human agencies, then the turbidity of the whole atmospheric column would probably have changed by at least the same relative amount (30% in 10 years) as that of the atmospheric column above observatory level. Now suppose that the doubling time of the human-derived dust load in the atmosphere is comparable to that of CO_2 emission, i.e. between 15 and 20 years based on the data of Revelle and Suess [12]. The rate of change of dust load implied by the Mauna Loa trend corresponds to a doubling time of somewhat less than 30 years. For these two estimates of doubling time to be consistent, about one half of the total dust loading of the atmosphere over Hawaii during the 1960s would have had to be derived from human activity. This leaves an improbably small remainder to be divided between volcanic dust, salt particles, and other natural aerosols known to be present in marine atmospheres.

Now let us assume (with Bryson) that this experience at Mauna Loa is representative of world-wide changes in turbidity. By a judicious combination of the Mauna Loa data with the estimates of relative world-wide abundances of volcanic and human-derived particulate loadings for the same period of time shown in Figure 3, we can arrive at a crude estimate of the contributions of each of these kinds of particulate loading to the *total* particulate load of the atmosphere. Let $\Delta T/T$ represent the relative change of dust loading over Hawaii (derived from the sealevel-reduced turbidity at Mauna Loa) before and after 1963, which we now attribute entirely to the arrival of volcanic dust from Agung (whence $\Delta T/T \simeq 0.13$). Further, let L represent the total atmospheric particulate load, V the part of L due to Agung volcanic dust, and H the part of L due to human-derived particulate loading. We then have the relation

$$\Delta V/L = \Delta T/T \simeq 0.13$$

where the Δ-symbols represent changes before and after 1963. This indicates that the total particulate load is roughly 8 times the incremental change of volcanic dust load due to Agung, which in turn corresponds closely to the total volcanic dust load of the atmosphere after 1963. Since according to Figure 3 the volcanic dust load after 1963 was of the order of 10 times the human-derived particulate loading, that is $\Delta V/H \simeq \simeq V/H \simeq 10$, we can also estimate that the contribution of human-derived particulate loading to the total particulate loading in the 1960s was less than 2%. Of greater interest than the total human-derived loading, however, is the *change* in human-derived loading during the 10-year period. If the data in Figure 3 are only very roughly correct, the change in H between 1957 and 1967 was less than half of the mean value of H in the period. Therefore the *change* in human-derived particulate loading in the 10-year period contributed less than 1% to the change of total atmospheric particulate loading during the same period. This is to be compared with the changes in total particulate loading, of the order of 10%, that are attributable to *volcanic* dust load changes in the same period.

If the foregoing analysis is only very roughly sound, we must conclude that although the total particulate load in the atmosphere due to human activity is approaching the average load due to volcanic activity (Figure 3), secular changes in the former have been a full order of magnitude slower in recent years than the irregular changes of volcanic dust loading. Thus, it seems reasonable to conclude that if the recent fluctuation of planetary temperature can be traced to planetary variations of atmospheric dust content, then volcanic activity rather than human activity is to be held primarily responsible for the temperature fluctuation.

B. ESTIMATION OF THE THERMAL EFFECT OF VARIABLE ATMOSPHERIC PARTICULATE LOADING DUE TO MAN'S ACTIVITIES

I have already remarked that various difficulties prevent us from making a reliable determination of the thermal effects of variable atmospheric particulate loading. This is true of all forms of particulate loading, both natural and human-derived. I do not intend to pursue this problem in any detail here. Rather, I propose merely to consider the estimated thermal effects of variable volcanic dust loading in the atmosphere and to derive from these a measure of the *upper limit* of the thermal effects of variable particulate loading due to human activities, that is consistent with the conclusions of the previous section.

A number of years ago I attempted to estimate the influence of major volcanic eruptions on the mean temperature of the earth [5]. The procedure was to determine the average temperature change following each eruption by means of a superposed epoch analysis of the same temperature data shown as Figure 1 of this paper (up to 1960). The interested reader is referred to the original paper for further details of the method. On this basis, it was estimated that an 'average' eruption (average severity index of about 2 on the scale of Table I) was followed by a 5-year-average temperature anomaly of the order of −0.1 °C. This estimate may be compared with the world-average temperature change of −0.3 °C since 1940, which was a period in which 6 eruptions of severity class 2 or greater occurred at an average interval of between 3 and 4 years. Only one of these latter-day eruptions (Mt. Spurr, 1953) was included in the earlier superposed epoch analysis, so that for all practical purposes the two figures are statistically independent. This comparison suggests that the magnitude of the recent world-wide cooling is more than double that which might have been expected from changes of volcanic dust loading of the atmosphere since 1940, but allowance for cumulative temperature effects that may apply when eruptions occur in relatively rapid sequence – as they did after 1940 – could conceivably resolve this difference. Thus we may tentatively attribute a major part (if not all) of the world-wide cooling trend after 1940 to increased volcanic dust loading of the atmosphere.

With this information we can arrive at a liberal estimate of the contribution of human-derived particulate loading to the temperature decline in recent years. Referring back to the discussion of the previous section, it was noted that the total human-derived particulate loading during the 1960s was nearly a full order of magnitude less than the post-Agung volcanic dust load. Even if we generously attribute all of the cooling since 1940 to post-Agung dust loading, we thus conclude that the *total* temperature effect of the human-derived particulate loading of the atmosphere in the 1960s was of the order of 0.05 °C. For a realistic doubling time of between 15 and 20 years for particulate loading from this source, we further conclude that *increases* of human-derived particulate loading of the atmosphere between 1940 and the present would have contributed no more than about 0.05 °C to the observed decline of temperature in the same period.

On the basis of these estimates, a temperature scale has been added to the margin of Figure 3. By reference to this scale, the atmospheric dust load chronology shown in the figure can be converted to a measure of its contribution to planetary temperature changes since 1850. I would, however, caution the reader not to interpret the figure too literally from this viewpoint, since crudely estimated temperature effects of roughly estimated dust loadings may add up to something very unlike reality.

6. Conclusions

On the basis of the foregoing analysis, it appears highly likely that the increase of carbon dioxide during the past century has been much more effective in altering planetary temperatures than has the concomitant increase of atmospheric particulate loading attributable to human activities. Even so, the warming influence of CO_2 has been sufficient to account for only about one third of the 0.6 °C rise of global-mean temperature observed between 1880 and 1940. The cooling influence of human-derived particulate loading is therefore likely to have been completely negligible during this warming period, but to have become somewhat more significant after 1940 when a

0.3 °C fall of global-mean temperature developed. The fact remains, however, that recent increases of human-derived particulate loading appear not to have been sufficient to offset the warming effect of further CO_2 increases since 1940. Thus it is very difficult to see how these particulate loading increases could have been the origin of the cooling trend observed after 1940. At the same time it is possible to attribute a major part of this latter-day cooling, if not all of it, to *natural* particulate loading changes that have accompanied a recent enhancement of world-wide volcanic activity. In this respect, it appears that man has been playing a very poor second fiddle to nature as a dust factory in recent years.

What are the implications of all this for the future? As I have already remarked, the warming influence of CO_2 has in the past exceeded the offsetting cooling effect of human-derived particulate loading. It is likely, however, that this imbalance between the two air pollution effects will not persist indefinitely. The reason for this is to be found in the relatively rapid doubling time of human-derived particulate loading. The doubling time of *accumulation* of fossil CO_2 in the atmosphere is apparently about 23 years. On the other hand, the doubling time of human-derived particulate loading is probably nearer 15 to 20 years, a value commensurate with the rate of *injection* of

TABLE II

Estimated future maximum global temperature anomaly due to combined thermal influence of carbon dioxide and particulate loading of atmosphere by human activities (for various assumed doubling times)

Assumed doubling time (years)		Maximum temperature anomaly (excess over temperature c. 1970)	
for CO_2	for particulate loading	Magnitude in °C (°F)	Date (A.D.)
23	20	> 10 (> 18)	> 2100
23	15	1.1 (1.9)	2040
	15*	0.3 (0.5)	2000
23	10	0.3 (0.5)	1985
	10*	0 (0)	1970

* Temperatures effect of particulate loading changes assumed double that estimated in text.

fossil CO_2 into the atmosphere as estimated by Revelle and Suess [12] and by Keeling [10]. If these dissimilar doubling rates are to be maintained in future decades, a time must eventually come when the cooling effect of particulate loading catches up with the warming effect of CO_2 and will surpass it thereafter. The date at which this cross-over will occur, and the maximum temperature anomaly due to the combined influence of both forms of air pollution which will be attained on that date, are shown for various assumptions in Table II.

Table II indicates that the cross-over date will probably arrive after 2000 A.D., following an additional warming of the order of 1 °C or more. In order for the cross-over date to arrive before 2000 A.D., the doubling time for particulate loading would have to be shortened to the rather unrealistic value of 10 years, or the magnitude of the thermal effect of particulate loading would have to exceed the magnitude estimated in this paper by a factor of two. In the latter event, the further warming would be limited to a few tenths of one degree, and the cross-over date (2000 A.D.) would be followed by an accelerating temperature decline as the particulate loading influence rapidly gains a decisive upper hand over the CO_2 warming. It is to be expected, of course, that natural climatic fluctuations will continue to occur in future decades – just as they have occurred in the past – and add considerable 'noise' to this tentative projection of our climatic future.

Acknowledgements

I am particularly grateful to H. H. Lamb for making available unpublished data on volcanic eruptions, that were indispensable in the construction of Table I; and to C. D. Keeling for illuminating discussions of the carbon dioxide problem.

References

[1] U.S. Public Health Service: 1962, *Air Over Cities*, SEC Tech. Report A62-5, Cincinnati; U.S. Public Health Service: 1969, *Air Quality Criteria for Particulate Matter*, NAPCA Publ. No. AP-49 U.S. Government Printing Office.
[2] Junge, C. E.: 1963, *Air Chemistry and Radioactivity*, Academic Press, N.Y., p. 228.
[3] Lamb, H. H. and Johnson, A. I.: 1959, *Geogr. Annaler* **41**, 94–134; 1961, *ibid.* **43**, 363–400.
[4] Mitchell, Jr., J. M.: 1965, in *The Quaternary of the United States* (ed. by H. E. Wright and D. G. Frey), Princeton University Press, pp. 881-901.
Mitchell, Jr., J. M.: 1967, in *Encyclopedia of Atmospheric Sciences and Astrogeology* (ed. by R. W. Fairbridge), Reinhold Publ. Co., pp. 211-213.
[5] Mitchell, Jr., J. M.: 1961, *Ann. N.Y. Acad. Sci.* **95**, 235–250;
Mitchell, Jr., J. M.: 1963, in *Changes of Climate, Arid Zone Research* **XX**, UNESCO, Paris, pp. 161–181.
[6] Scherhag, R.: in *Berliner Wetterkarte* (Berlin, annual issues 1965, 1966, 1967).
[7] Emiliani, C. and Geiss, J.: 1967, *Geol. Rundschau* **46**, 576–601.
[8] Goldberg, E.: this volume, p. 178.
[9] Callendar, G. S.: 1958, *Tellus* **10**, 243–248; C. E. Junge in [2].
[10] Keeling, C. D.: personal communication.
[11] Pales, J. C. and Keeling, C. D.: 1965, *J. Geophys. Res.* **70**, 6053–6075.
Brown, C. W. and Keeling, C. D.: 1965, *J. Geophys. Res.* **70**, 6077–6085.
[12] Revelle, R. and Suess, H. E.: 1957, *Tellus* **9**, 18–27.
[13] Bolin, B. and Eriksson, E.: 1959, in *The Atmosphere and Sea in Motion* (ed. by B. Bolin), Rockefeller/Oxford Press, pp. 130–142; C. E. Junge in [2].
[14] Manabe, S. and Strickler, R. F.: 1964, *J. Atmos. Sci.* **21**, 361–385.
Manabe, S. and Wetherald, R. T.: 1967, *J. Atmos. Sci.* **24**, 241–259.
[15] Lamb, H. H.: personal communication.
[16] Dyer, A. J. and Hicks, B. B.: 1968, *Quart. J. Roy. Meteorol. Soc.* **94**, 545–554; C. E. Junge in [2].
[17] Davitaia, F. F.: 1965, *Trans. Soviet Acad. Sci.*, Geogr. Ser., No. 2.
[18] Peterson, J. T. and Bryson, R. A.: 1968, *Science* **162**, 120–121.

For Further Reading

1. On the nature and significance of recent fluctuations of climate: *Changes of Climate*, Arid Zone Research XX, UNESCO, Paris, 1963.
2. On the nature, origin, distribution and properties of gaseous and particulate constituents of the atmosphere:
   C. E. Junge, *Air Chemistry and Radioactivity*, Academic Press, New York, 1963.
3. On the effects of atmospheric particulate matter on solar radiation and climate:
   U.S. Public Health Service, *Air Quality Criteria for Particulate Matter*, U.S. Government Printing Office, Washington, D.C. (in press). Preliminary edition available from National Air Pollution Control Administration, Washington, D.C., 1969.
4. On the optical and other meteorological effects of volcanic dust in the atmosphere:
   W. J. Humphreys, *Physics of the Air*, 3rd ed., McGraw-Hill Book Company, New York, 1940.

CLOUDINESS AND THE RADIATIVE, CONVECTIVE EQUILIBRIUM

SYUKURO MANABE

Geophysical Fluid Dynamics Laboratory/ESSA, Princeton, N.J., U.S.A.

Abstract. The dependence of the temperature of the earth's surface upon the cloud cover at various altitudes is estimated. The effect of contrail on the surface temperature is discussed.

Professor Bryson (p. 136) suggests that contrails may have a significant effect on the climate of the earth's surface. Professor Schaefer (p. 158) speculates that the increase of the atmospheric turbidity caused by human activity could affect the number of ice crystal nuclei and accordingly, the cloudiness. With these thoughts in mind, I would like to discuss how clouds affect the temperature of the earth's surface.

As you know, clouds have the following two radiative effects:

(1) They reflect solar radiation.

(2) They decrease the outgoing terrestrial radiation at the top of the atmosphere (because the temperature of cloud top is usually colder than that of the earth's surface).

The first effect lowers while the second effect raises the temperature of the earth's surface.

In order to evaluate the effect of clouds on the temperature of the earth's surface quantitatively, Manabe and Wetherald (1967) performed a series of computations of the radiative, convective equilibrium of the atmosphere with various distributions of cloudiness and with a given distribution of relative humidity. For our study, the reflectivity of solar radiation was assumed to be 20% for cirrus clouds, 48% for middle clouds (As), and 69% for low clouds based upon the study of Haurwitz (1948). For the computation of the flux of terrestrial radiation, low and middle clouds were assumed to be completely black, while cirrus clouds were assumed to be half black (according to Kuhn's recent measurements which show that cirrus are far from black). Since I described how we obtained the radiative, convective equilibrium earlier (p. 25), I shall not explain it any further.

Figure 1 shows the dependence of the equilibrium temperature of the earth's surface upon cloud cover. According to this figure, an increase of low clouds lowers the temperature of the earth's surface markedly because they have a large reflectivity and a relatively warm emission temperature. On the other hand, an increase of high clouds could raise the surface temperature because they have a low reflectivity and a low emission temperature. It is noteworthy that the increasing of low clouds by 3% has a comparable effect with the halving of the CO_2 content in lowering the equilibrium temperature of the earth's surface. (Refer to the discussion on p. 145 concerning the effect of CO_2-content on the climate.)

According to the estimate of Professor Bryson contrails cover about 0.8% of the sky. If contrails have the same optical characteristics as Mr. Wetherald and I have

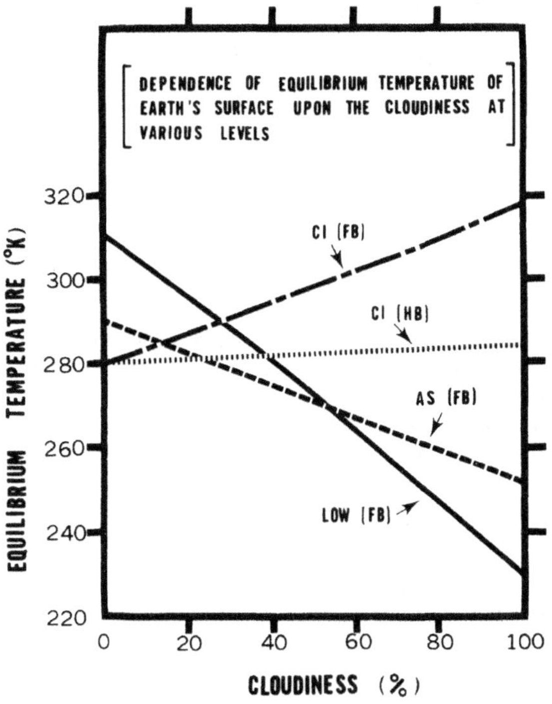

Fig. 1. Radiative, convective equilibrium temperature at the earth's surface as a function of cloudiness. (Cirrus, altostratus, low cloud.) FB and HB refer to full black and half black for terrestrial radiation respectively. (By Manabe and Wetherald, 1967.)

assumed for cirrus, then contrail covering 1% of the sky would have a negligible effect upon the temperature of the earth's surface. But if they are full black for the terrestrial radiation, 1% sky coverage of contrails would raise the equilibrium temperature by about .3 °C and have a marginally significant effect. Our results are also highly dependent upon the reflectivity of solar radiation assumed for the computation. Therefore, it is necessary to establish the optical characteristics of ice clouds such as cirrus and contrails in order to determine whether or not the radiative effects of contrails are significant in modifying the climate of the earth.

References

Kuhn, P. M.: private communication.

Haurwitz, B.: 1948, 'Insolation in Relation to Cloud Type', *J. Meteor.* **5**, 110–113.

Manabe, S. and Wetherald, R. T.: 1967, 'Thermal Equilibrium of the Atmosphere with a Given Distribution of Relative Humidity', *J. Atmos. Sci.* **24**, 241–259.

THE INADVERTENT MODIFICATION OF THE ATMOSPHERE BY AIR POLLUTION

VINCENT J. SCHAEFER

Atmospheric Sciences Research Center, State University of New York at Albany, N.Y., U.S.A.

Abstract. The subject of my concern at the present time as it relates to global pollution is an aspect of the problem which may affect our welfare sooner than the more obvious deterioration of our environment as it relates to health, esthetics or property damage.

While it is quite possible that a disaster directly related to the build-up of particulate matter and gases in the atmosphere may occur within the next decade unless some major improvements are achieved toward reducing the rapid increase in air pollution now prevalent, I believe that only those already in poor health will be the immediate victims. After all – if one considers that the cigarette smoker insults his lungs with a concentration of particulate matter of at least 10 million particles per cm^3, – which is more than ten thousand times greater than the concentration of particles in country air and more than a hundred times worse than the air of a badly polluted city – one can realize that the human body can stand a considerable amount of physiological abuse. This is not to say we should push our bodies to their limit of compensatory abilities.

However, the phenomenon which concerns me most at the present time is the distinct probability that air pollution has already begun to inadvertently modify the atmosphere in which we live and the climate on which our complex civilization is based.

It is this subject that I would like to discuss with you in this Symposium.

1. Introduction

There has been a very noticeable increase in air pollution during the past ten years over and downwind of the several large metropolitan areas of the U.S.A. such as the Northwest – Vancouver-Seattle-Tacoma-Portland; the West Coast from San Francisco-Sacramento-Fresno-Los Angeles; the Front Range of the Rockies from Boulder-Denver-Colorado Springs-Pueblo; the Midwest – Omaha-Kansas City-St. Louis-Memphis; the Great Lakes area of Chicago-Detroit-Cleveland-Buffalo; and the Northeast – Washington-Philadelphia-New York-Boston. The worst accumulation of particulate matter occurs at the top of the inversion which commonly intensifies at night at levels ranging from 1000 to 4000 feet or so above the ground. This dense concentration of air-suspended particles is most apparent to air travellers. Thus, it has not as yet disturbed the general public except during periods of stagnant weather systems when the concentration of heavily polluted air extends downward and engulfs them on the highways, at their homes and in their working areas.

2. Recent Modification of Our Air Environment

Until recently there is little question that except in very exceptional cases, natural processes dominated the mesoscale weather systems by initiating the precipitation mechanism. The effluent from the larger cities was quickly diluted by the surrounding 'country air' so that at a distance a few miles downwind of a city, little evidence of air pollution could be detected.

The recent spread of urban developments due to better roads and the massive proliferation of people and automobiles has led to a nationwide network of county, state, and interstate highways. This interconnection of thousands of smaller towns with large cities and the phenomenal increase in auto, truck and air traffic has caused a massive reduction in the regions which have 'country' type air. This increase in massive air contamination is of fairly recent origin. It is not easy to document this fact in the detail I would prefer since we have not had reliable automatic recording equipment for measuring Aitken, cloud and ice nuclei until the last few years. However, using simpler devices with which we made measurements at a number of scattered locations during the past 10 years, we have in the past year used the same techniques to make comparative observations. The measurements indicate an increase in airborne particulates at these sites of at least an order of magnitude during this 10 year period. At Yellowstone Park in the wintertime, which has the cleanest air we have found in the continental United States, the background levels of Aitken nuclei have increased from less than 100 to the 800–1000 per milliliter range within a 5-year span. At Flagstaff, Arizona where in 1962 the background levels ranged from 100–300 the concentration now lies between 800–3000. At Schenectady, New York, the average concentration of these nuclei has risen from less than 1000 to more than 5000 with values occasionally exceeding 50000 per milliliter.

While it is difficult to ascribe these increases to any one cause, it is obvious that the increased demand for electric power, the large increase in garbage and trash incineration and the automobile, are likely to represent the major sources of increased pollution, especially since many industrial plants have been forced to reduce their pollution due to more rigorous regulations.

Just as it is not easy to place the blame for increased air pollution specifically on the power plants, incinerators and automobiles, it is equally difficult to demonstrate clean cut or unequivocal atmospheric modification to these sources. I am confident that in time there will be ample proof of these effects which are now inadvertently modifying the atmosphere.

The presence of high concentrations of visible as well as invisible particulates above and downwind of our cities produces a heat island effect as real as a sun-drenched Arizona desert or a semitropical island in the Carribean.

Those cities like Boston, New York and Philadelphia which are not affected by geographic barriers as are Los Angeles, Salt Lake City, or Denver are able to get rid of much of their effluent whenever the wind blows. Their plumes of airborne dirt extend as visible streamers for many miles downwind of the source areas. In the case of the metropolitan New York City-northeastern New Jersey complex, these plumes will be found in the upper Hudson Valley, in southeastern New England or over the Atlantic Ocean.

Commercial airline pilots flying the Atlantic are often able to pick up these pollution plumes hundreds of miles at sea. Hogan recently obtained data which provide a quantitative measurement of the New York effluents near the surface of the Atlantic between the U.S.A. and Europe. This same paper [1] amply demonstrates a similar zone of air pollutants being exhuded over the seas surrounding Europe, the British Isles and the east and west coast of the United States.

3. Properties of Maritime and 'Country' Air

We have known for 20 years that maritime air is characterized by low levels of both cloud droplet and Aitken nuclei. Vonnegut showed [2] by a very simple experimental device that about 50 effective nuclei at lower water saturation droplet formation existed on the upwind coast of Puerto Rico where the trade wind clouds are seen. We were all much surprised when we established the nature of trade wind clouds during our research flights near Puerto Rico in 1948 [3]. Following these activities, I pointed out [4] the large difference noticeable even then between the 'raininess' of the clouds upwind of the island and those which formed over the land after entraining the polluted air from San Juan, the sugar fields and refineries, the cement mills and the myriads of charcoal pits which dotted the island, each sending out its plume of bluish smoke. In our studies in the vicinity of Puerto Rico we observed that in many instances trade wind clouds would start raining by the time the clouds had a vertical thickness of not more than a mile while those over or immediately downwind of the island often reached three times that thickness without raining.

During a continent-wide flight over a large area of Africa I found [5] an even more spectacular effect of inadvertent cloud seeding. As a result of the massive bush and forest burning initiated by the inhabitants preceding the onset of the rainy season, huge cumulus clouds, some of them reaching a height of more than 35000 feet (vertical thickness 4–5 miles) were observed which were not producing any rain. Instead the clouds grew so high that very extensive ice crystal plumes hundreds of miles long extended downwind of the convective clouds. No evidence of glaciation was observed in the side turrets of the clouds indicating a deficiency of ice nuclei at temperatures warmer than the homogeneous nucleation temperature of $-40\,°C$. Thus it appeared that the precipitation process was being controlled almost entirely by coalescence and that so many cloud droplet nuclei were being entrained into the clouds from the fires below, that the coalescence process was impaired so that no rain developed. If ice nuclei were present, they were probably deactivated by the high concentration of smoke particles and gases flowing into the base of the clouds. Similar effects have been observed on a smaller scale in the Hawaiian Islands. During the trade wind cloud regime, clouds which form over sugar cane fields when they are burned prior to harvest are actually larger than the surrounding clouds but they have never been observed to rain even though smaller ones nearby produce showers. Warner more recently has documented such observations [6].

A further observation of secular change in the microphysics of clouds has been observed in the vicinity of large cities during airplane flights through convective clouds. The observations I have noted in particular were made in commercial twin engine planes over the past 10 years. Of recent years it has been noticed that such clouds often have so many cloud droplets in them that visibility is restricted so much that the engine is hardly visible. In my earlier observations I can never recall being in clouds so opaque that the wing tips could not be seen. Several of my colleagues have reported similar experiences.

Perhaps the most impressive field evidence of inadvertent weather modification is the overseeding of supercooled clouds which is readily observed over and downwind of our northern cities in the wintertime.

4. Ice Crystals from Polluted Air

Although I have been observing such phenomena for more than 10 years, the effect was brought to my attention in a vivid way during a flight from Albany to Buffalo on December 20, 1965. After flying above a fairly thin deck of supercooled stratus clouds downwind of the Adirondack Mountains, I noted a massive area of ice crystals above and downwind of Rochester, New York. The crystals were so dense that the reflection from the undersun* was dazzling as illustrated in Figure 1. Since that time

Fig 1. Undersun photographed downwind of Rochester, New York December 20, 1965.

* The undersun is an optical phenomenon caused by the specular reflection of the sun from the surfaces of myriads of hexagonal plate ice crystals. In order to produce an undersun, it is necessary for the crystals to consist of smooth-surfaced plates which float with their long axes horizontal to the ground. They thus act as many tiny mirrors. If the crystals were not hexagonal plates but rather prisms, the optical effects would include under parhelia and other reflections which are well known and have been related to crystal types during our winter studies at Yellowstone Park.

I have observed similar high concentrations of crystals at low level above and downwind of most of the large northeastern cities such as New York, Albany, Utica, Syracuse, and Buffalo as well as Detroit, Chicago, Sacramento and Los Angeles. In all instances the ice crystals were observed at low level (below 5000 feet above the ground in most instances), and extending for at least 50 miles downwind of the city sources and without cirrus clouds above the areas affected. In a few instances when the plane passed through the crystal area, I observed the particles to be like snow dust, though in a number of instances after landing I observed very symmetrical though tiny hexagonal crystals drifting down from the sky.

5. Misty Rain and Dust-Like Snow

For the past several years I have also been observing a number of strange snow and rain storms in the Capital District area in the east central part of New York State. These storms consist of extremely small precipitation particles. When in the form of snow, the particles are like dust having cross sections ranging from 0.02 cm (200 μ) to 0.06 cm (600 μ). When in the form of droplets, they often are even smaller in diameter, at times being so tiny that they drift rather than fall toward the earth. When collected on clean plastic sheets, the precipitation is found to consist of badly

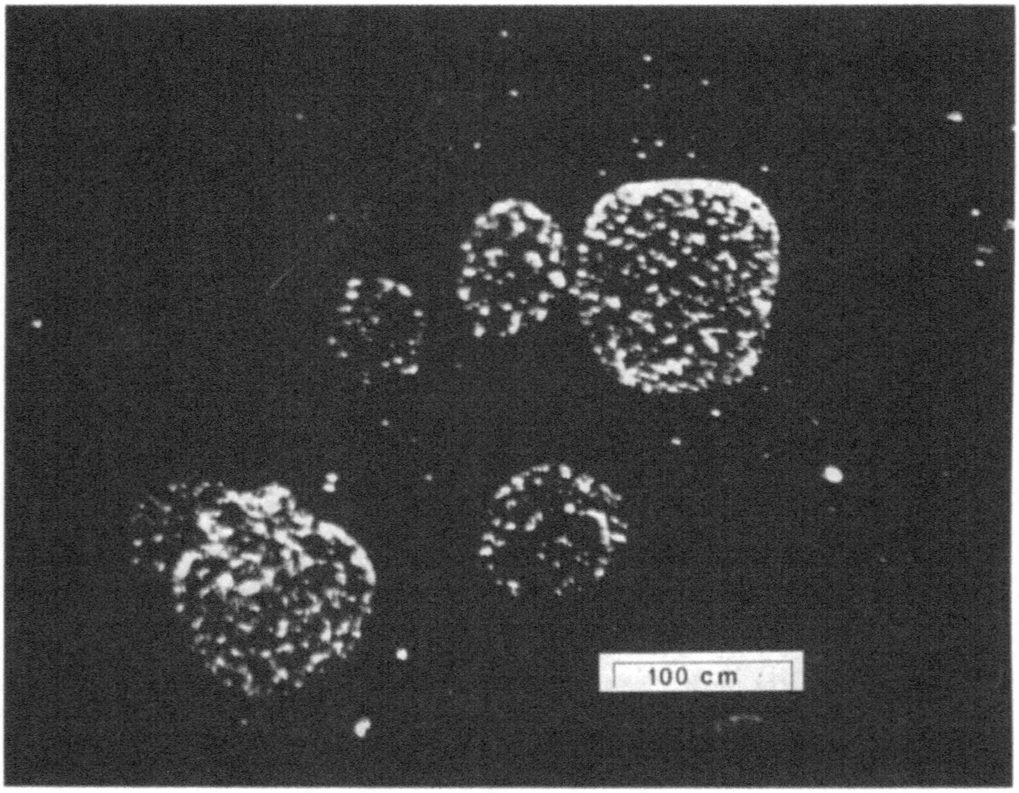

Fig. 2. Residue remaining on glass slide after polluted, misty rain droplets have evaporated.

polluted water. It is a well-known fact that precipitation 'cleanses the air'. In the past much of this cleansing action has been ascribed to the sweeping up of suspended aerosols by rain and snow. Little attention has been given to the possibility that submicroscopic particulates from man-made pollution may in fact be initiating and controlling precipitation in a *primary manner* rather than being involved in the secondary process wherein precipitation elements coming from 'natural' mechanisms serve to remove the particles by diffusion, collision and similar scavenging processes.

My first evidence that there might be substances in urban air which would react with other chemicals was encountered while studying ice nucleation effects at the General Electric Research Laboratory in 1946 [7]. At that time I found that laboratory air contained aerosols which would react with iodine vapor to form very effective ice nuclei but that when the air was free of particulate matter, no further ice particles would form.

6. Potential Ice Nuclei from Auto Exhaust

In 1966 I published a paper [8] which suggested that air pollution in the form of automobile exhaust could account for the high concentration of ice crystals which I have observed downwind of the larger cities in the United States and in any area where a considerable number of automobiles are used. My laboratory studies have shown that submicroscopic particles of lead compounds produced from the combustion of leaded gasoline can be found at concentrations exceeding 1000 per cm^3 in auto exhaust. These were measured by exposing auto exhaust samples to a trace of iodine vapor before or after putting the samples into a cold chamber operating at $-20\,°C$. Presumably this reaction with iodine formed lead iodide which is an effective sublimation nucleus for ice crystal formation. Evidence that the active ingredient in auto exhaust consists of submicroscopic particles of lead was determined by comparing its temperature ice nucleation activity pattern with that of lead oxide smoke produced by electrically sparking lead electrodes which was also reacted with a trace of iodine vapor. Figure 3 is a photomicrograph of replicated ice crystals formed on submicroscopic lead compounds from auto exhaust reacted with iodine vapor. One of the problems related to the evidence that leaded gasoline is responsible for the ice crystals observed in laboratory and field experiments concerned with auto exhaust is the source of the iodine needed to produce the lead iodide reaction. All evidence thus far encountered shows that only a few hundred molecules of iodine are required to produce a nucleating zone for ice crystal formation. The amount of iodine reported in oceanic [9] air (the order of 0.5 U G. per m^3) is orders of magnitude greater than would be required to activate such particles.

I have recently completed further studies in Arizona, New York and France [10, 11] and have found that wood smoke and other organic sources add iodine to the air which could react with the auto exhaust submicroscopic lead compounds which are always present in urban pollution. Hogan has recently showed [12] that similar reactions will proceed from the vapor phase.

Admittedly we are dealing with chemical reactions in the realm of surface and even

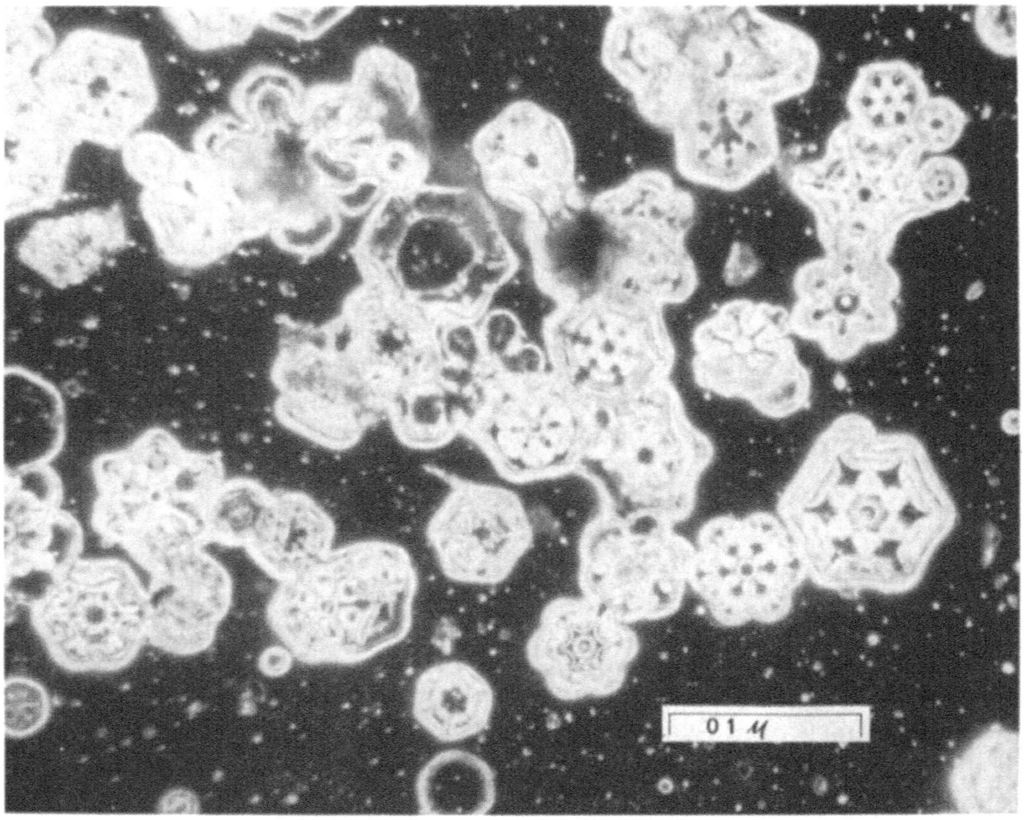

Fig. 3. Photomicrograph of replicated ice crystals formed at −16 °C in a supercooled cloud on nuclei from auto exhaust exposed to trace of iodine vapor.

'point' chemistry as Langmuir termed such molecule by molecule reactions. This is an area of particulate research for which there is very little experimental data or practical experience. The size of the primary lead particles from auto exhaust which are 0.008–0.010 μ diameter are far too small for analysis by any currently available chemical reaction techniques. All of my laboratory experiments indicate that the submicroscopic particles in auto exhaust which react with iodine vapor act only as nuclei for ice formation from the vapor phase. No evidence has been found that they act as freezing nuclei.

7. The Effect of Large Concentrations of Ice Crystals

The presence of high concentrations of tiny ice crystals in air colder than 0 °C over thousands of cubic miles raises interesting aspects of the dynamics of weather systems. Such crystals continually modify small supercooled clouds soon after they form. The net result is a reduction in the number of local rain or snow showers and the production of extensive sheets of 'false' cirrus. Bryson has pointed out [13] that cirrus sheets and even the presence on a large scale of airborne dust exerts a measurable decrease of insolation. If a much larger supply of moist air moves into such a region, the entrainment of high concentrations of crystals by more vigorous supercooled clouds may trigger the formation of a massive storm through the release of the latent heat of sublimation. Langmuir described [14] such a storm system which he believed was initiated and then intensified when dry ice in successive seeding operations was put into the lower level of a rapidly developing storm.

8. Findings of Project Air Sample

In order to determine whether or not polluted air above cities contained particles which would react with free iodine molecules, eight transcontinental flights have been made under our auspices during the Fall of 1966 and 1967 and the Spring and Fall of 1968. The flight routes are shown in Figure 4. A Piper Aztec aircraft was fitted with instruments which could measure in a semi-quantitative manner the concentration of atmospheric particulates which would become ice nuclei by the reaction with iodine, and which would also measure natural nuclei for ice crystal formation. The iodine reactions were conducted in a cold chamber at $-20\,^{\circ}\mathrm{C}$. The determination of naturally occurring nuclei was done at $-22\,^{\circ}\mathrm{C}$. In addition, measurements were also made of Aitken nuclei (a measure of polluted air) and cloud nuclei. This last measurement which is made at very low water saturation is also a measure of the degree of

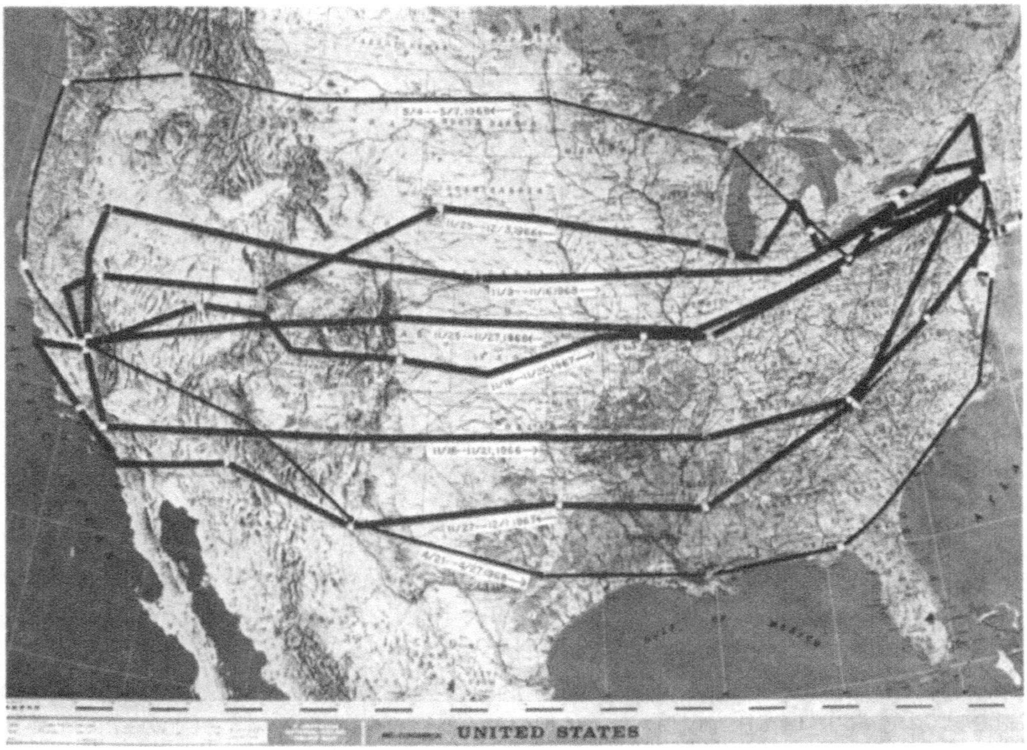

Fig. 4. Flight routes of Project Air Sample 1–4 1966–1968.

air pollution since values above about 50 cloud nuclei and 500 Aitken nuclei per cm^3 is indicative of some degree of air pollution. The flight samples were made mostly just below the top of the haze layer which ranged from 1500 to 5000 feet above the ground throughout the flights. Figure 5 illustrates the aircraft used in Project Air Sample flights and the A.S.R.C. Experimental Laboratory. Of the 266 measurements in November 1966, 31 were made on the ground. All of these showed

Figure 5. Experimental Laboratory for Cloud Physics and aircraft used for air pollution studies.

excessive pollution levels. Great care was exercised in making these observations to avoid contamination from the engine exhaust of the aircraft being used for the measurements.

At several locations observations were made above as well as within the upper part of the haze layer. In every instance the air above the visible top of the haze layer was low in lead particles while that just below the top or farther down showed very high concentrations.

All other locations where counts of the ice nuclei were low involved regions free of pollution sources. Of the 266 observations 108 or 40% of the measurements were in areas such as upwind of cities (9); above large lakes (8); above haze layers (22); and woods and farms (33). The 60% remaining had values of potential ice nuclei of 100 per liter or more. Some 115, all of them above or downwind of cities had values in excess of 200 per liter which I consider would lead to definite overseeding of the atmosphere with ice particles if suitable moisture was available. Values of 1000 per liter or more occurred at 101 of the stations. If concentrations of ice crystals that high occurred, the cloud would resemble a stable ice fog such as occurs at Fairbanks, Alaska or the Old Faithful area of Yellowstone Park [15] when the temperatures are colder than −40 °C. With crystal concentrations of this magnitude, the particles grow very slowly if at all and thus remain floating in the air for extended periods. This then reduces the incoming solar radiation to a noticeable degree. If such areas are extensive, they cannot help but cause changes in the weather patterns of the affected areas.

Similar findings characterized our second, third and fourth round-trip transcontinental flights covering more than 25000 additional miles and consisting of over 1500 more observations. In practically every instance where polluted air was present, high values in potential ice nuclei (using the iodine reaction) was found. The only exceptions were instances where the plumes of steel mills, forest fires and other highly concentrated effluents were measured in areas where auto exhausts could contribute very little if anything to the sampled air.

Fig. 6. Example of pollution which occurs along Lake Erie at Buffalo, New York, November 20, 1967.

Fig. 7. Snow showers falling from thin clouds about 80 miles downwind of Buffalo, New York, November 20, 1967.

Figure 6 illustrates the type of pollution which still occurs along Lake Erie at Buffalo and Figure 7 a zone of snow falling from low clouds about 80 miles downwind of Buffalo at a location where the iodine-activated nuclei had dropped from 5000 per liter measured at Buffalo to 500 near Cayuga Lake and the Aitken count from 25000 per milliliter to 4500 as measured at 3000 feet above the terrain.

Figure 8 illustrates the fantastic amount of smoke which shrouded the metropolitan New York area on November 23, 1966 when one of the first dangerous smog alerts was sounded by New York City health officials. Just prior to obtaining this picture the airplane was flown up through the smog. At 600 feet the cloud nucleus count was 2000 per milliliter, the Aitken count 25000 and the ice nuclei measured were 0 for the natural background and 50000 to 100000 per liter for ice nuclei activated with iodine vapor.

Figure 9 shows the conditions at Albany, New York, on the previous day. At 1200 feet the cloud droplet nuclei numbered 900 per milliliter, the Aitken count was 4000, the natural ice nucleus background was 0 but the concentration of ice nuclei activated with iodine was 50000 per liter. These are concentrations which are commonly observed in the air below the top of the inversion over and downwind of all large

Fig. 8. Example of severe pollution over New York City, November 23, 1966.

cities. Although in many instances the smoke concentrations in such areas are not as spectacular as shown in Figure 7 and 8 since a large portion of the smog particles tend to be submicroscopic, such areas are generally characterized by a veil of brownish gray haze.

9. Flight Observations of Inadvertent Seeding

It is quite feasible to detect and observe the massive systems of ice crystal nuclei which produce inadvertent effects on cloud and weather systems due to man's activities. This is accomplished most easily by riding on the sunny side of a jet aircraft.

I observed and photographed three such systems in 1967 during a flight from Buffalo, New York to Denver, Colorado by way of Chicago, returning directly from Denver to New York City.

A. ICE CRYSTALS RELATED TO POLLUTED AIR

On Wednesday, December 6, 1967, I left Buffalo at 1035 by Boeing 727 landing at Detroit and Chicago. Upon take-off I noted a heavy pollution pall over Buffalo extending westward to the horizon. Just west of Buffalo we climbed above stratiform

Fig. 9. Example of heavy air pollution at Albany, New York, November 22, 1966.

clouds estimated to be at 15000 feet or lower which consisted of very high concentrations of ice crystals as established by an undersun. This extensive zone of ice crystals was observed all the way to Detroit and was associated with visibly polluted air. We flew at 20000 feet where the temperature was −20 °C. Enroute from Detroit to Chicago I found the same condition to exist from the 1108 take-off until 1132 at which time only supercooled clouds were visible. At the same time all evidence of polluted air disappeared, visibility between cloud decks was unlimited and no further trace of ice crystals could be seen as we landed at Chicago. The air pollution from Chicago was being carried to the southeast over Indiana about thirty miles south of our jet route.

B. ICE CRYSTALS PRODUCED BY DUST FROM PLOWED LAND

Upon take-off at Chicago in a Boeing 707 at 1320 CST, the air was clear, several decks of stratiform clouds were visible with no evidence of ice crystals. Heading west I saw no ice crystals until 1424. Just previous to that time a peculiar zone of dusty air could be seen ahead of us extending toward the southwest. Within a few minutes a brilliant undersun could be seen which persisted for the next half hour. When we finally emerged from the affected zone over northeastern Colorado it was quite obvious that the 300 miles long zone of ice crystals was due to very extensive dust storms caused by 50–100 mph katabatic ground winds pouring down out of the Front Range of the Rockies and blowing top soil from the extensive wheat fields extending from north-eastern Colorado to the region about 50 miles east of Pikes Peak. The low level dust was rising only from tilled land, the grassy areas such as the Pawnee Grass Lands were unaffected.

A similar massive dust storm which produced very extensive cloud seeding was observed by me on the afternoon of April 12, 1967 between Amarillo, Texas and Denver, Colorado. This affected region was so extensive and had such a profound effect on the Great Plains and midwest weather systems that I was able to identify it and see its effects over western Illinois two days later.

On the return flight from Denver on December 8, a third source of inadvertent weather modification was observed. Take-off in a DC9 occurred at 1206 MST on a non-stop jet flight to New York City. Very fine snow was falling at the ground upon take-off. Four minutes afterward we climbed above an extensive area of ice crystals. A bright undersun became visible and was seen continuously all the way from the Denver area to the Atlantic Ocean east of New Jersey. Jet contrails appeared to be the source of these crystals throughout the entire flight which was conducted at 37000 feet. More than a dozen different planes were seen coming from the east within the flight corridor we were using, most of them several thousand feet below us. From time to time we were close to contrails being made by planes at our level but ahead of us.

The most striking effect observed as the sharp line of demarcation between the area affected by contrail seeding along our flight corridor and an extensive area of high altocumulus cloud (or cirrocumulus) which paralleled our zone at its southern extremity. This region of non-modified clouds was estimated to be about 10 to 20 miles away and extended over large regions of the country. I expect that an effect such as was observed could be seen on satellite cloud photographs.

Perhaps the most disturbing feature about inadvertent weather modification is that in a subtle manner it seems to be changing the nature of clouds over increasingly large areas of the globe. Much of our current consideration of cloud seeding assumes the ubiquity of supercooled clouds and the effectiveness of a seeding material for triggering the instability of such systems.

If pollution sources lead to increased dustiness from ill-used land, more cloud nuclei from burning trash and many more ice nuclei from the lead-permeated exhaust of internal combustion engines, not only will we lose the possible advantage we now have of extracting some additional water from our sky rivers, but we might even be confronted with a drastic change in our climatological patterns.

Interesting climatological evidence of inadvertent weather modification was found by Chagnon [16] to exist in the area downwind of the Chicago, Illinois-Gary, Indiana complex of extensive urban, highway and steel mill concentrations.

A very noticeable increase in precipitation and storminess is evident in the records of the past three decades. The LaPorte, Indiana region whose record is cited as evidence of this effect is downwind of the heavy pollution source mentioned above as well as the close proximity to a very moist air source in the form of Lake Michigan. It is a common observation to see a lake effect street of cumulus clouds extending in the convergence zone south and southeast of Lake Michigan. The combination of very moist air and an abundance of ice nuclei are apparently in very favorable juxtaposition for an optimum reaction to occur. The LaPorte anomaly was first observed by a local weather observer which was then evaluated by Chagnon. He found that there has been a notable increase in precipitation starting about 1925 with definite increases since that time also of the number of rainy days, thunderstorms and hail storms. There has been a 31% increase in precipitation, 38% of thunderstorms and 240% increase of hail incidences. The increases show a marked correlation with the production of steel.

Since these data were obtained entirely from an evaluation of the climatological records, it is of great importance that careful 'on-the-spot' field observations should be made in the LaPorte area to establish the atmospheric dynamics which are responsible for the apparent change in the precipitation pattern of that area. It is particularly important that the concentration of particulate matter be correlated with storm patterns. The weather systems at the mesocale level should especially be studied to determine whether the area receiving increased precipitation is in the center or edge of the city-industrial plume effluent and the properties of the moist air moving in from Lake Michigan.

10. Experimental Production of Large Areas of Ice Crystals

During the past 10 winters field operations have been developed by our Yellowstone Expeditions in which we have established certain relationships of ice crystal concentrations in the free atmosphere. The early morning inversions of the Old Faithful Geyser Basin in the wintertime often have liquid water contents ranging from 0.5 to 1 G/m^3. This rich supply of moisture is contained within a strong ground-based inversion having a vertical thickness of about 100 meters. At a distance of 2000 meters from a point source of seeding, ice crystal concentrations up to 10000 per liter have been measured. Such crystals at $-12\,^{\circ}$C are hexagonal plates with cross sections of from 10 to 1000 μ, the size depending on concentration and moisture supply. Those of 200 μ occur typically at a concentration of 200 per liter with a fall velocity of 10 cm per sec. The brilliance of the undersun and related optical phenomena indicates that the number of crystals observed in areas caused by air pollution, jet contrails or dust storms often have concentrations as high or higher than observed in our experiments. Thus at Yellowstone we have an ideal outdoor laboratory to study some of the factors which must be better understood if we are to work out the physical interactions resulting from the inadvertent modification of the atmosphere.

11. The Need and Opportunity to Study These Phenomena

The effects cited are but a few examples of many which I have observed and photographed during the past few years. It is the rule rather than the exception that such massive zones of ice crystals can be observed over large areas of the country which can be related to man-caused modification.

Such occurrences must be exercising a detectable effect on the weather systems of the northern hemisphere. I feel that nowhere near enough effort has been directed toward the establishment of an organized and continuing study to determine the effect of such inadvertent seeding mechanisms on the synoptic weather patterns of our country. Such studies should have a major place in the World Weather Watch and the Global Atmospheric Research Project. I strongly recommend that the part played by atmospheric particulates should become an important research feature of this program.

There is a critical need for knowledgeable field scientists having an extremely broad scientific background who can work effectively in the real atmosphere under all types of conditions and extract quantitative and meaningful data from such systems.

Our Universities must place far more emphasis on this type of training than is being done at present. The eventual understanding of these complex interrelationships do depend on computers, electron microscopes, mass spectrometers and other costly instruments and equipment. However, the real atmosphere is the thing that must be understood and it is not enough to rely on data obtained by automatic instruments and uninformed field men as is too often the case. It is not easy to conduct efficient field operations. We must approach nature to an ever increasing degree but this confrontation must involve 'intelligent eyes', an understanding of the physics, chemistry and electricity of the reactions which can occur and a zeal to understand the things which combine to produce atmospheric phenomena.

References

[1] Hogan, A.: 1968, 'Experiments with Aitken Counters in Maritime Atmosphere', *J. Rech Atmos.* **3**, 53.
[2] Vonnegut, B.: 1950, 'Continuous Recording Condensation Nuclei Meter', *Proc. First Natl. Air Pollution Symposium*, Pasadena, Cal. **1**, 36.
[3] Schaefer, V. J.: 1953, 'Final Report Project Cirrus Part 1 Laboratory, Field and Flight Experiments', Report No. RL-785 General Electric Research Laboratory, Schenectady, N.Y.
[4] Schaefer, V. J.: 1956, 'Artificially Induced Precipitation and its Potentialities', in *Man's Role in Changing the Face of the Earth* (ed. by W. L. Thomas), University of Chicago Press.
[5] Schaefer, V. J.: 1958, 'Cloud Explorations over Africa'. *Trans. N.Y. Acad. Sci.* **20**, 535.
[6] Warner, J.: 1968, 'A Reduction in Rainfall Associated with Smoke from Sugar Cane Fires – An Inadvertent Weather Modification', *J. Appl. Met.* **7**, 247–251.
[7] Schaefer, V. J.: 1948, 'The Production of Clouds Containing Supercooled Water Droplets or Ice Crystals Under Laboratory Conditions', *Bull. Am. Met. Soc.* **29**, 175.
[8] Schaefer, V. J.: 1966, 'Ice Nuclei from Automobile Exhaust and Iodine Vapor', *Science* **154**, 1555.
[9] Junge, C. E.: 1963, 'Air Chemistry and Radioactivity', Academic Press, New York.
[10] Schaefer, V. J.: 1968, 'Ice Nuclei from Auto Exhaust and Organic Vapors', *J. Appl. Met.* **7**, 113.
[11] Schaefer, V. J.: 1968, 'The Effect of a Trace of Iodine on Ice Nucleation Measurements', *J. Rech. Atmos.* **3**, 181.
[12] Hogan, A.: 1967, 'Ice Nuclei from Direct Reaction of Iodine Vapor with Vapors from Leaded Gasoline', *Science* **158**, 800.

[13] Bryson, R. A.: 'Is Man Changing the Climate of the Earth?', *Saturday Review* p. 52, April 1, 1967.
[14] Langmuir, I.: 1962, 'Results of the Seeding of Cumulus Clouds in New Mexico,' *The Collected Works of Irving Langmuir*, Pergamon Press, New York, Vol. II, pp. 245–262.
[15] Schaefer, V. J.: 1962, 'Condensed Water in the Free Atmosphere in Air Colder than $-40\,°C$', *J. Appl. Met.* **1**, 481.
[16] Chagnon, S. A.: 1968, 'LaPorte Weather Anomaly, Fact or Fiction', *Bull. Amer. Met. Soc.* **49**, 4.

PART IV

WORLDWIDE OCEAN POLLUTION BY TOXIC WASTES

INTRODUCTION

The oceans have been termed the 'ultimate sink' for the natural wastes of the world. Most of these are carried there by rivers. But, with the increase of technology, new types of wastes are being generated, some of which, like lead, add a considerable load to the naturally occurring content; others, like DDT, are a completely foreign substance to the ocean environment. Incidentally, both lead and DDT are partially transported to the oceans by the atmosphere.

Goldberg traces the routes of introduction, and catalogs the amounts of exotic materials which are introduced through pollution. The United States' contribution to the world total is remarkably high, and both rates are increasing. New kinds of chemicals are added almost daily, and radioactive discharges will undoubtedly rise. The fate of the pollutants is no longer under control. "The solution to pollution is dilution!" becomes the slogan. The actual fate is quite complicated and may involve slow chemical reactions or uptake by organisms.

Biologically-active materials released into the environment, whether on land, in lakes, or oceans, lead to general ecological effects conforming to well-known patterns. These are described by Woodwell, with particular reference to DDT.

Marine pollution has also quite direct implications on the harvest of food from the sea. As related by Ketchum, this problem has become especially pronounced in estuaries which are more subject to pollution than the open ocean. It will become even more important when we try to enhance food yields through aquaculture in the coastal waters.

Finally, the ocean is an all-pervasive influence on the environment of the globe and plays a central role in the interaction of all ecosystems. Lundholm describes some of these interactions – the transfer of pollutants between atmosphere and hydrosphere.

THE CHEMICAL INVASION OF THE OCEANS BY MAN

EDWARD D. GOLDBERG

Scripps Institution of Oceanography, University of California, San Diego, La Jolla, Calif., U.S.A.

Abstract. The alteration of the chemical composition of the oceans by man is proceeding at an ever increasing rate due to rises in population and in industrialization over the world. In some instances the activities of society are increasing the levels of substances in sea water; in others, materials foreign to the oceanic realm are being introduced.

An understanding of how the oceans may be modified can be found in considerations of the oceans as a chemical system. The characteristics of marine chemical reactions – the sites of the processes, their time constants, and the nature of the reactants – are considered along with the routes of transport of materials from the continents to the oceans.

1. Introduction

Man, a land organism, is influencing the chemical composition of sea water more than any of the species that live within the marine environment. The world ocean receives the discharges of an ever increasing world population in ever increasing amounts and in ever increasing complexities. Nearly all materials used by man in his social, agricultural and industrial activities are not retained by him but are dispersed, often in degraded forms, about the surface of the earth. The oceans by intent of man or by their very expanse receive a major portion of these materials. Such metabolic wastes of our civilization can increase the levels of chemical species, already a part of the oceanic domain, or they may contain substances alien to the marine environment.

2. Mercury

Lead and mercury are two elements whose entries into the oceans from the continents appear to be about equally influenced by normal weathering processes and by the discharges of civilization. The transfer of mercury from the continents to the oceans via the rivers is of the order of 5000 tons per year. About one-half of the world production of mercury (9200 tons per year, *Minerals Year Book*, 1967a) is utilized by agriculture and industry with a subsequent release to the environment. Since the early decades of this century organic mercury compounds have been used as fungicides, primarily as seed dressings and as catalysts in the production of such chemicals as chlorine and sodium carbonate. Such usages are held responsible for the increases in the mercury contents of birds collected in Sweden where the levels in the 1940s and 1950s were 10 to 20 times those from previous years (Berg *et al.*, 1966) and may explain the covariance of high concentrations of atmospheric mercury and smog in the San Francisco Bay region (Williston, 1968).

Between 4000 and 5000 tons of mercury per year most probably enter the oceans as a result of the release of man utilized compounds to the rivers and to the atmosphere. As yet this input has not been sought or identified in the marine environment. If the

Singer (ed.), Global Effects of Environmental Pollution.

oceanic organisms behave as do their fresh water counterparts, the fish of the sea will accumulate both organic and inorganic mercury compounds directly from solution with high concentration factors. On the other hand, such materials are not taken up to an appreciable degree by plants (Hannerz, 1968).

3. Lead

Man appears to be responsible for an input of lead into the oceans equal to that of natural processes. In the northern hemisphere about 350000 metric tons of lead, as the anti-knock agent lead tetra-ethyl, are burned in automobile internal combustion engines and subsequently introduced into the atmosphere. Tatsumoto and Patterson (1963) suggest that about 250000 metric tons of lead are annually washed out over the oceans and about 100000 metric tons over the continents in correspondence to their relative areas. Some of the land fallout eventually reaches the oceans as river runoff. This compares with an annual input of lead into the oceans through natural weathering processes of 150000 metric tons. This impingement by man has raised the average lead content in surface waters of the northern hemispheric oceans from about 0.01–0.02 to 0.07 micrograms/kg of sea water in the 45 years since the introduction of lead as an anti-knock chemical (Chow and Patterson, 1966).

4. Gases

The rate of atmospheric introduction of such carbon containing substances as carbon monoxide, carbon dioxide and low molecular weight hydrocarbons is approaching, and may have already exceeded, the rate of production of organic matter, about 10^{16} g of carbon per year (Steeman Nielsen, 1960), by the photosynthesizing plants of the sea (Table I). While society's contribution of such materials increases with time, the natural production by plants is presumed to remain more or less constant. The U.S.A. in comparison with the rest of the world appears to be responsible for around one-third to one-half of such waste products.

All of the materials injected into the atmosphere do not enter the marine environ-

TABLE I

Estimated rates of injection of materials into the atmosphere (g/year)

Material	U.S. rate	World rate
CO	7×10^{13}	2×10^{14}
Sulphur oxides	2×10^{13}	8×10^{13}
Hydrocarbons	2×10^{13}	8×10^{13}
Nitrogen oxides	8×10^{12}	5×10^{13}
CO_2		9×10^{15}
Smoke particles	1×10^{13}	2×10^{13}

ment. Some, existing as stable aerosols, can be assumed to enter the ocean in precipitation relative to an entry on the lands in proportion to the relative areas of ocean and land under the atmospheric system being considered. This concept was used in the input of lead aerosols and may well define the additions of fly ash, insecticide carriers such as talc, and debris from nuclear detonations to the oceans. For substances existing in the gaseous phases, the relative importance of sea water to other sinks for their removal from the atmosphere usually must be evaluated on an individual basis. For example, carbon monoxide appears to be oxidized rapidly to carbon dioxide soon after its injection into the atmosphere and perhaps only a small percentage of its production ever dissolves in sea water.

5. Chlorinated Hydrocarbons

Substances not naturally produced in the marine environment also are being discharged by the world society. Perhaps the most abundant of the synthetic pollutants is p, p′ DDE, a degradation production of p, p′ DDT which is the principal component of technical DDT (Risebrough, personal communication). Although the DDE does not possess the pesticidal toxicity of DDT, it does induce liver enzymes in mammals, birds and fish which hydroxylate and thereby cause an increase in solubility of a wide variety of non-polar compounds, especially hormones. Other chlorinated hydrocarbons such as dieldrin, endrin, heptachlor epoxide, and benzene hexachlorides have been found in various members of the marine biosphere and are distinguished by both their chemical stability and high chlorine contents. As yet these substances have not been quantitatively assayed in sea water itself; however, the available data indicate that marine fish and birds already contain as much, if not more, pesticide residues than their fresh water counterparts.

Another group of halogen-containing organic materials, the polychlorinated biphenyls (PCB), are widely dispersed in the marine biosphere, being present in most of the world's sea birds. According to Risebrough (personal communication) they may be the second most important pollutant in the seas. The weight ratio of DDT to PCB in sea birds of the Pacific appears to be between 5 and 10, with individuals from industrial areas having higher PCB levels than those from more remote regions. Used in the plastics, paint and rubber industries, they have effects similar to the previously mentioned pesticide residues as inducers of the steroid hydroxylate enzymes.

6. Radioactivity

Radioactive species, released to the oceans from the recent nuclear device testing programs of the U.S.A., U.S.S.R., U.K., China and France may be found in all oceans and in all members of the marine biosphere. Such programs have introduced such novel nuclides as radioactive Sr^{90} and Cs^{137} into the world's seas, as well as substantially increasing the amounts of such cosmic-ray produced nuclear species as C-14 and H-3 in the surface layers of the ocean and the atmosphere. With these radioactivities man has left his imprint in all parts of the world ocean which covers two-thirds of the earth's surface to an average depth of 4 km. Up to the present time there is no evidence that these pollutants have had any widespread adverse effect upon marine communities. On the other hand, their very presence has stimulated much governmental support of research which has markedly broadened our oceanographic knowledge.

7. Hydrocarbons

Some discharges from our society, less notorious than the emission of automobile exhausts, industrial chimneys and power plants, may eventually end up in the oceans and significantly affect their compositions. For example the gasoline lost solely by evaporative processes, both during transfer from refinery to truck, from truck to holding tank, or from holding tank to automobile and during standing in reservoirs or carburetors, represents $2\frac{1}{2}\%$ by volume of the total annual production (5×10^{11} liters for 1965 in the U.S.A. – *Minerals Year Book*, 1967b), or 1.0×10^{12} grams. The fate of these gasoline hydrocarbons is as yet unknown, but it is conceivable that they do contribute to the dispersed organic species of surface waters.

Similarly, the evaporation of dry cleaning solvents, such as the widely used tetrachloroethylene (perchlorethylene) takes place at a rate of 3.6×10^{11} grams per year in the United States. The chemical stability of such chlorinated hydrocarbons, as used in the dry cleaning industry, suggests that a substantial portion of the evaporative loss will be absorbed by the oceanic system.

8. Marine Chemistry

The characteristics of natural processes that have been studied within the oceanic system provide an important background for predicting the fates of materials introduced by man. Many marine chemistries can be described in the following ways:

(1) The reactants are in extremely low concentrations – micromolar and below.

(2) The periods needed to produce discernible amounts of materials are measured in geologic units, hundreds to millions of years, in comparison to the time constants usually involved in laboratory experiments, ranging from fractions of a second to days.

(3) The reactions take place at phase discontinuities – the atmosphere-ocean, the atmosphere-sediment, and the atmosphere-biosphere.

The formation of the ferromanganese minerals, accumulations of iron and manganese oxides, occurs in all oceans and at all depths and illustrates these three characteristics. The reactants in these sedimentary processes, iron and manganese, have sea water concentrations of 2×10^{-7} and 4×10^{-8} molar respectively, yielding the minerals in which their contents achieve levels of 15 to 20% by weight on the average.

In some areas of the oceans they may cover the entire sea floor (Menard, 1964). Their patchy distribution precludes an average distribution figure, but it is of interest to note that, in the southwestern Pacific, a region with 26–46% of the bottom paved with manganese nodules has been found, while in an area in the North Pacific, centered at a latitude 18–20 °N possesses a bottom coverage of the minerals between 10 and 100%.

The chemical reactions involved in the formation of the minerals involve most probably the oxidation of dissolved manganese in sea water. A simple and direct representation of the process is given by

$$Mn^{++} + 2OH^- + \tfrac{1}{2}O_2 = MnO_2 + H_2O\,.$$

This reaction probably occurs heterogeneously with a solid surface which is a good absorber for oxygen, converting molecular oxygen to atomic oxygen. Such substances as MnO_2 or Fe_2O_3 may provide the sites for the reaction as these oxides have lower states of oxidation. Thus, they can, in principle, give up oxygen forming metal ions with lower oxidation states as has been pointed out by Duke (1967).

The ferromanganese minerals accrete at rates of millimeters per million years, or, perhaps in more dramatic units, at rates of atomic layers per year. This growth, measured by several independent radioactive techniques reflects one of the slowest chemical reaction rates ever measured.

The remarkable accumulation by members of the marine biosphere of many metals which exist in sub-micromolar concentrations is amply recorded in the oceanographic literature. The quite remarkable observation of Carlisle (1958) that certain tunicates concentrate vanadium (concentration 2×10^{-8} molar), and others its vertical periodic table neighbor niobium (concentration 10^{-10} molar), and still others neither element, but none that can accumulate both elements, emphasizes the most unusual specificities that have been encountered in recent investigations. Perhaps the comment of Nichols *et al.* (1960) that "for any given chemical element there will be found at least one planktonic species capable of spectacularly concentrating it" will direct attention to the range of possibilities for the uptake by organisms of unusual chemicals introduced by our advancing civilization.

9. Residence Times

The relative reactivities of elements in the marine environment can be measured by the lengths of time they spend in the ocean, subsequent to introduction, and before precipitation to the sea floor. This period, the so-called residence time, can be calculated under the assumption that the world's oceans are a single, well-mixed reservoir of water at steady state. The input of substances from the rivers is complemented by an equal amount of material accommodated in the sediments; thus, the composition of the oceans is taken to be invariant with time.

This schematization of the oceans yields residence times for the alkalis and alkaline earths of the order of millions to hundreds of millions of years, emphasizing their lack of reactivity in solution. Other elements such as aluminum, iron, the rare earths and thorium have much shorter residence times, hundreds to thousands of years, periods of the order of, or less than, that of mixing times for oceanic water masses. These elements, in part, enter or exist in the oceans as particulate phases and rapidly settle to the sedimentary deposits. Some are the reactants in the chemical reactions that form such authigenic substances as the ferromanganese minerals, zeolites, etc. Their entry into the oceans as solids and/or their high chemical reactivity can account for their low residence times.

Thus, the chemical invasion of the oceans by man involves time periods quite different from those encountered in terrestrial water systems. Where a river renews itself annually, lakes in periods of decades or centuries, the oceanic environment maintains its components for times orders of magnitude greater. Present-day introductions of materials will be measurable for many, many thousands of years in the future.

10. Transport into the Ocean

The distributions of materials introduced from the continents to the oceans are influenced by the transporting agency: rivers, winds, or glaciers. Man has imposed two additional paths: discharges from ships and sewage outfalls. Of consequence are the long distances of transport that takes place through natural processes, readily seen by a consideration of mineral distributions in the sediments. Especially diagnostic are the clay minerals which constitute about 60% by weight of the non-biological phases of the deposits. Four clay minerals are of importance. Each can characterize a geological or climatic regime. Illite is a degradation product arising from the alteration of crustal rocks in temperature climates. The impress of the South American rivers upon the sediments of the western Atlantic Ocean is clearly demonstrated by gradients in illite contents. Wind-transported illite from the Asian deserts by the jet streams causes unique compositional patterns of this mineral in the North Pacific.

Chlorite has sources in the abundant low grade metamorphic rocks, sedimentary shales and argillites of the Arctic and Antarctic regions, solids which undergo little or no chemical change during weathering. The transport of chlorite can take place through glacial movement as can be defined by the decreasing gradients going from higher to lower latitudes. Also, chlorite concentrations outline the extent of the influence of the St. Lawrence and Alaskan rivers upon the Atlantic and Gulf of Alaska, respectively.

Montmorillonite can be characterized as the clay mineral indicative of a volcanic regime although there are inputs directly from the continents. In the South Pacific and South Atlantic the highest concentrations are in mid-oceanic areas, with gradually decreasing concentration gradients going toward the continents. Often in association with the montmorillonite are phillipsite and glass shards, remnants of a previous volcanic activity, presumably within the ocean basins.

The formation of kaolinite occurs in geological regimes undergoing intense weathering, such as the low latitude regions of the earth. The high kaolinite abundances are in general confined to the equatorial sediments. River-transported kaolinite is seen influencing the composition of sediments off the Congo and Niger rivers in the Atlantic, while wind-transported kaolinite has been postulated to explain the decreasing kaolinite contents in the deposits off western Australia and in the equatorial regions of the Atlantic off the African coast.

Clearly, we can predict in part the dissemination of materials within the marine environment, knowing the transporting agency that delivers them from a continental area. It was not unexpected to find pesticides, applied to the agricultural areas in Africa and transported 6000 km across the Atlantic by the Northeast Trades, in dust collected in islands of the Caribbean.

11. Consequences of Human Activities

Finally, and of crucial import, is the need for an understanding of the consequences of man's chemical invasion of the oceans. Can this intrusion affect the composition of the marine communities by altering the mortality rates of one or more species? Can certain areas be made inappropriate as habitats for creatures of the sea as a result of the introduction of unpleasant substances? What is the probability of any undesirable substances injected by man returned to him in his food products recovered from the marine environment?

A substantial portion of the scholarly activity of our society has been dedicated to an understanding of its surroundings. In part these studies have been motivated by curiosity; in part by the need to protect man's institutions against the ravages of nature such as winds, rains and flood waters. The ability of mankind to significantly alter the characteristics of the earth's surface emphasizes a third basis for environmental studies. The responses to the man-imposed changes in nature can be detrimental and perhaps there can result the loss or the restricted uses of a valuable resource. The ability to predict undesirable results along these lines may lead to necessary and protective policies concerning such encroachments.

References

Berg, W., Johnels, A., Sjostrand, B., and Westermark, T.: 1966, *Oikos* **17**, 71–83.

Bureau of Mines: 1967a, *Minerals Year Book*, Volume 1, 1966.

Bureau of Mines: 1967b, *Minerals Year Book*, Volume 1, 1966, p. 810.

Carlisle, D. B.: 1958, *Nature, London* **181**, 922.

Chow, T. J. and Patterson, C. C.: 1966, 'Concentration Profiles of Barium and Lead in Atlantic Waters off Bermuda', *Earth Planetary Sci. Letters* **1**, 397–400.

Duke, F.:1967, 'Factors Determining Chemical Oxidation and Reduction in Solution', in *Principles and Applications of Water Chemistry* (ed. by S. D. Faust and J. V. Hunter), Wiley, pp. 370–379.

Hannerz, L.: 1968, 'Experimental Investigations on the Accumulation of Mercury in Water Organisms', *Institute of Freshwater Research, Drottningholm* **48**, 120–176.

Menard, H. W.: 1964, *Marine Geology of the Pacific*, McGraw-Hill New York.

Nichols, G. D., Curl, H., and Bowen U.T.: 1960, 'Spectrographic Analyses of Marine Plankton', *Limnol. Oceanogr.* **4**, 472–478.

Risebrough, R.: personal communication.

Steeman Nielsen, E.: 1960, 'Productivity of the Oceans', *Ann. Rev. Plant. Physiol.* **11**, 341–361.

Tatsumoto, M. and Patterson, C. C.: 1963, 'The Concentration of Common Lead in Sea Water', in *Earth Science and Meteoritics* (ed. by J. Geiss and E. D. Goldberg), North-Holland Publ. Co., pp. 74–89.

Williston, S. H.: 1968, 'Mercury in the Atmosphere', *J. Geophys. Res.* **73**, 7151–7155.

For Further Reading

1. E. D. Goldberg, 'Chemistry in the Ocean', in *Oceanography*, AAAS, Washington, D.C., 1961, pp. 583–97.
2. L. G. Sillen, 'The Ocean as a Chemical System', *Science* 156 (1967), 1189–97.

CHANGES IN THE CHEMISTRY OF THE OCEANS: THE PATTERN OF EFFECTS*

GEORGE M. WOODWELL

Biology Dept., Brookhaven National Laboratory, Upton, N.Y., U.S.A.

Abstract. The changes in the chemistry of the oceans reported by Dr. Goldberg can be expected to reduce the structure of plant and animal communities according to well-known patterns. The changes in structure will be similar to those observed under conditions of accelerated eutrophication in fresh water lakes. One of the most conspicuous changes involves loss of highly specialized top carnivores; there are ample signs that avian carnivores are disappearing now from oceanic communities. The present losses are due to accumulations of persistent toxic compounds, especially DDT. Evidence suggests that at current rates of use DDT residues will continue to accumulate in the next decades to levels 2 to 3 times current levels, affecting even oceanic fisheries.

Dr. Goldberg has shown conclusively that the chemistry of the oceans is changing due to the activities of man. The changes are many, they are occurring very rapidly if measured by the time required for the evolution of life, and their effects, although profound in some instances, are not easily observed. We might ask what can be said about the general pattern of these effects on living systems and their significance.

Prediction of the specific effects on natural communities of any single change in environment is usually difficult and sometimes impossible, but the broad pattern of changes brought by drastic changes in environment such as those that Dr. Goldberg has outlined are predictable. They are similar on land and in water; similar in all natural communities.

The basic pattern is one of simplification: progressive reduction of the structure of natural communities as the disturbance becomes more severe; a shortening of food chains; elimination of top carnivores and a shift toward larger numbers of a few kinds of plants and animals that are small and have rapid rates of reproduction. The pattern is familiar on land in places subject to frequent disturbances such as roadsides, unstable soils, and even cultivated lands, excluding of course the crops. In such places plants tend to be low in stature, rapidly growing, and to have rapid, often asexual, reproduction as well as other characteristics that make them particularly successful under adverse conditions. In water the problems of accelerated eutrophication provide an example that is becoming too familiar. Here enrichment with nutrients causes rapid growth of certain small plants, ultimately changing the characteristics of the water body completely. The complex food webs that in fresh water once supported trout and salmon are lost, replaced by short food chains where the consumers are plant or detritus eaters such as carp or mullet or, in the worst situations, simply anaerobic organisms of the decay food chains. Lake Erie has become the classical example although there are many others. Accelerated eutrophication of the type that

* Research carried out at Brookhaven National Laboratory under the auspices of the U.S. Atomic Energy Commission.

Dr. Goldberg's data suggest for the oceans may be accompanied by accumulation of toxic substances speeding a shift to anaerobic conditions, becoming unquestionably 'pollution'.

That disturbance should cause this pattern of change is hardly surprising. The course of biological evolution is in the opposite direction, tending with time to develop greater numbers of species using the diverse and continually growing resources of environment with an ever higher degree of specialization. The specialization that accompanies the development of a diversity of species and a community that has a complex structure makes the community vulnerable to disturbance. If the disturbance is severe enough or continued, the specialists are replaced by other species that can survive the changing conditions.

There are many ways of measuring details of the structure and function of communities. One of the most comprehensive involves measurement of exchanges of energy. It is comprehensive in that it integrates details of structure and function, showing the quantitative relationships between populations. It is convenient for our purposes because it shows what occurs when the structure of any ecosystem is lost.

The energy driving natural communities is transferred from the green plants to other populations according to simple rules. For instance, it appears that only about 10% of the energy entering the plant population is available for consumption by herbivores; 10% of that entering the herbivores is available to the first level of carnivores; and so on through two, perhaps three, levels of carnivores. Clearly, there is a quantitative relationship between populations of different trophic levels. Equally clearly, populations of highly specialized carnivores at the top of the trophic structure are at greatest hazard if the structure is disturbed and we should look to them for clues that the structure is being disturbed.

It is less clear that such a trophic structure is also vulnerable to toxic substances that are concentrated by trophic-level effects. The concentration occurs because the rate of supply of the substance through the food chain exceeds the rate of loss. Losses may include chemical breakdown as well as excretion. When successive links in a food chain concentrate a substance, concentration factors of thousands or even hundreds of thousands above environmental levels are possible, putting carnivores, and top carnivores in particular, at special hazard.

It is not surprising to discover that some of the earliest signs that the structure of natural communities is being lost appear as reductions in populations of carnivores and especially of carnivorous birds. Birds of course have high metabolic rates and therefore consume large amounts of food in proportion to their own weights. They are also conspicuous and populations of rare or unusual birds, including many of the top carnivores, have been watched very closely for many years by both amateur and professional ornithologists. Changes in the populations, especially drastic reductions, are quickly recognized.

Such signs are now available for the oceans. One of the best known is the abrupt decline in the reproductive success of the Bermuda petrel which was first observed by the Bermudan naturalist, David Wingate. After considerable thought and experimentation he finally guessed, based on evidence from other carnivorous birds around the world, that the cause might be the accumulation of DDT residues. The concentrations of residues carried by the birds were later shown to be in the range known to affect reproductive success in other species, suggesting strongly that the Bermuda petrel, which never comes into direct contact with man or with areas sprayed with DDT, but rather feeds in the open ocean, is being affected by DDT residues that are passed along to it through the food webs of the North Atlantic. This observation of course means that the food webs are carrying significant burdens of DDT and leads one to guess that toxic effects must exist not only among carnivorous and scavenging birds, which are conspicuous and well known, but also among diverse aquatic animals, including even the oceanic fisheries. This conclusion is substantiated by Dr. Goldberg's data.

The question of what further changes will occur if current uses of persistent pesticides are continued is a real one. I have shown elsewhere that if DDT residues have a half-life in the environment of 10 years, which seems a minimum, and the annual release of DDT into the biosphere is 200 million pounds, then the equilibrium concentration will be approached only after 70 years. The residues circulating at the moment must be less than one-half the residues that will be circulating when chemical degradation precisely balances the releases of DDT, if current levels of use are maintained. There is every reason, then, to expect that DDT alone will account for a significant degradation of oceanic ecosystems, including the oceanic fisheries, in the next decades unless its use in places where it can contaminate the living systems of the earth is halted.

It would be a strange contradiction indeed if the agricultural and industrial interests that defend the use of DDT so recklessly on the grounds that it is needed for production of food for the earth's hungry millions, were to continue to be successful and in their success eliminate the oceanic fisheries. Adding to the irony would be the fact that these fisheries, already threatened by over-exploitation and by elimination of marshes and other hazards, offer at present almost the only short-term system for recycling the nutrient elements harvested in agriculture and dumped through civilization into our sewers and the sea. Unless harvested as fish, these nutrients are lost.

The increased burden of DDT is of course but one of the chemical changes that is occurring in the oceans. It is, however, clear now that the plant and animal communities of the oceans are being degraded as a result of the changes in oceanic chemistry caused by man. The pattern of change is clear: it is the pattern of eutrophication and pollution similar to that well known in Lake Erie and in thousands of other water bodies around the world and similar in broad outline to the changes that occur on land. We may be able to lose lakes in this way, even one of the Great Lakes, but is very doubtful indeed whether we can afford to lose the oceans.

The solution to such world-wide pollution problems lies not simply in controlling the pollutants: it lies of course there. But it lies equally importantly in providing a general context within which there is not strong, potentially overwhelming, pressure to use the earth so intensively as to pollute it. Such a context requires that resources be large in proportion to demands made on them. That most desirable, perhaps even essential, condition can be established only if both population and the incursions on environment made by technology are limited. If pollution problems are to be controlled in the long run, not merely mitigated temporarily, and aggravated in the long run, then it must quickly become the policy of nations to limit population and to restrict those aspects of technology that degrade the common resources, including air, water and land. The changes that Dr. Goldberg reports now in the oceans prove that nations can scarcely move rapidly enough in this direction.

General References

Hasler, A. D.: 1969, 'Cultural Eutrophication is Reversible', *Bioscience* **19**, 425–431.

Woodwell, G. M.: 1967, 'Radiation and the Patterns of Nature', *Science* **156**, 461–470.

Woodwell, G. M.: 1967, 'Toxic Substances and Ecological Cycles', *Scientific American* **216**, 24–31.

Woodwell, G. M.: 1969, 'Radioactivity and Fallout: The Model Pollution', in *Proc. of Rockefeller University/N.Y. Botanical Garden Symposium, Challenge for Survival*, April 25–26, 1968, Columbia University Press, in press.

Wurster, C. F. and Wingate, D. B.: 1968, 'DDT Residues and Declining Reproduction in the Bermuda Petrel', *Science* **156**, 979–981.

For Further Reading

1. *On the significance of pesticides residues to the earth's biota:*

   a. Miller, M. W. and Berg, G. C. (eds.): *Chemical Fallout: Current Research on Persistent Pesticides*, Chas. C. Thomas, Publisher, Springfield, Illinois, 531 pp.

   b. Moore, N. W. (ed.): 1966, *J. Applied Ecology* **3** (supplement), 155 pp.

2. *On the structure and function of ecosystems:*

   Odum, E. P.: 1970, *Fundamentals of Ecology'*, W. B. Saunders Co., 1970 (3rd edition in press).

3. *On the changes in the structure of natural communities caused by pollution:*

   Woodwell, G. M.: 1970, 'Effects of Pollution on the Structure and Physiology of Ecosystems', *Science* (in press).

BIOLOGICAL IMPLICATIONS OF GLOBAL MARINE POLLUTION

BOSTWICK H. KETCHUM

Environmental and Systematic Biology, National Science Foundation, U.S.A.

Abstract. The sea is discussed as an excellent source of animal protein, the lack of which contributes to malnutrition of large numbers of the human population. Although the sea produces large amounts of vegetable material, its harvest is unrealistic because it is so sparsely distributed. Consequently, the land must continue to produce the energy rich plant materials in the human diet. Pollution has already limited our harvest of seafood in estuaries, and the remaining areas, suitable for marine life, must be protected from additional pollution if we are to maintain and increase our harvest of protein from the sea. The potential supply, and its limitation, are briefly discussed and it is suggested that development of modern aquaculture would be desirable, in addition to the search for and exploitation of additional natural populations.

Dr. Goldberg's paper (p. 178) has amply demonstrated the fact that man's activity is now adding strange and unusual materials to the oceans on a global scale. His data also indicate that we, as residents of the U.S.A., are contributing in a major way to the gradual accumulation of this pollution in the oceans. Dr. Goldberg concluded with a statement that the most important effects of pollution are on the cycle of life in the sea. Dr. Woodwell has already expanded on the subject of these effects, particularly the effects of increasing pesticide levels (p. 186). I would like to discuss in more general terms the value of our marine living resources and the need to protect them if man is to continue to enjoy the harvest of the sea.

For centuries, man's pollution has ultimately reached the sea via the rivers and estuaries. It is probably true that, if the ocean were devoid of life, man could continue to pollute the oceans indefinitely with little fear of consequences since the volume available to dilute pollutants is enormous. The focus of the problem is necessarily a concern about the effects of our addition of pollutants on the organisms of the sea.

The question which will, no doubt, occur first to most people concerns our harvest of marine products and the effects of pollution on edible species. What is the chance of the return to man of some toxic product in the seafood he eats? More indirectly, what effect will our pollution of the sea have upon the availability of marine food products? The answer to both of these questions is obvious in the estuaries, where man's influence is greatest. Without adequate controls, infectious hepatitis from shellfish could be a serious health problem. In spite of our best efforts for the treatment of domestic sewage, large productive areas are closed for the harvest of shellfish for human consumption because of the danger of disease. Many of our estuaries are devoid today of those fish which once were abundant. The run of shad in the Hudson River is not as great as it was when I was a boy, and the Atlantic salmon has disappeared from most of our coastal rivers. Our coastal estuaries play a very important role, not only as a habitat for living organisms which grow and are harvested within their boundaries, but also as a breeding ground for many species of fish which we capture in offshore coastal waters. The remaining estuaries which are still clean enough to provide for the support of these natural resources must be protected if we are to continue to enjoy the harvest of the oceans. However, the ultimate and probably more serious effects of pollution, are much more subtle and involve the whole chain of events which culminate in the production of desirable species.

How important are these marine products to mankind? Dr. Byerly has stateb that we can produce the food needed for the expected human populations of the future (p. 104). He can rightly be proud of the record of the U.S. Department of Agriculture and of the American farmer for the dramatic and continued increase in the productivity of our farmland. I do not question his predictions but I would like to discuss three facts which were not mentioned during the discussion of terrestrial pollution and food production problems.

In the first place, the high production of food by agriculture requires the affluent society which we enjoy. It is not fertilization alone which has produced these high yields of agriculture. Cheap and abundant sources of energy are also necessary. The American farmer with his electrified farm and barn, with his tractors and combines, expends about 5 calories of energy for each calorie of food energy he produces. The energy used for these purposes today is mostly in the form of fossil fuel and contributes to the carbon dioxide problem which was the subject of one section of this symposium. The flow of energy, however, follows a one way street and, once degraded, it is not recoverable. The pre-industrial man cycled energy on a short time basis and essentially at a rate which could easily be maintained indefinitely by the process of photosynthesis. Thus his fuel for heat was wood which can be replaced in about 50 years in the forest and the horse power he used in plowing and harvesting his fields was derived from pasture lands which are annually reproduced. Today we are utilizing energy at a rate much faster than it can be reproduced by photosynthesis and, since we are living on an isolated planet, 'spaceship Earth', this process cannot go on forever. Today, of course, nuclear energy has offered the promise of an almost limitless supply of energy so that we have postponed the day of reckoning, perhaps forever. It should be clear, however, that we cannot vastly increase the productivity of the developing nations without providing them with an enormous supply of energy which they can utilize in order to match our agricultural productivity.

The second fact is that the methods used in our temperature zone agriculture cannot be used without extensive modification in tropical areas. Most of the hungry peoplet of the world are concentrated in the tropics and their agricultural production must be increased in order to ameliorate their problems of malnutrition and hunger. Dr. Commoner has discussed the decrease in organic nitrogen in our cultivated soils and the accumulation of nitrates in flowing water, ground water and in some foods (p. 70). In the tropics, the nutrient cycles are much more frugal and similar methods of agriculture would deplete the soils there at a greatly accelerated rate. Without the freezing winters, the decomposition of organic materials proceeds at a high rate throughout the year and tropical agriculture will require many precautions which we have considered unimportant in temperate regions for the past several generations.

The third fact that I would like to mention is that adequate nutrition of the world population requires more than food calories alone. Those people who depend on a single vegatable crop for their calories inevitably suffer protein deficiencies, either in total protein or because of a lack of one of the ten amino acids essential for human nutrition. The development of highly productive strains of grain and rice has frequently been accompanied by a decrease in the protein content. Approximately 1.5 billion persons, largely in tropical and subtropical areas, live on limited, protein deficient diets. Protein deficiency is a serious nutritional defect even among some of our own population, and it is truly serious among many people throughout the world.

Protein deficiency can be readily overcome by the addition of comparatively small amounts of animal protein to the diet of undernourished peoples. The harvest from the sea is a protein-rich product and it is even now supplying half of the annual intake of animal protein for nearly half of the world's population. This may seem incredible to Americans, since only about 5% of our own protein supply comes from the sea and, of this small fraction, about 60% is used indirectly as protein supplements to chicken and livestock feed. For the protein deficient nations, a daily supplement of 10 to 20 g of animal protein would suffice to overcome the debilitating effects of protein deficiency. The type of animal protein is of little consequence; seafood will serve just as well as beef, pork or chicken.

Thus, to maintain adequate nutritional balance, the products of the sea are already vital to half of the world population. Even though the animal protein provides only a small fraction of the calories needed for subsistence, this source of animal protein must be protected and increased if we are to have not only well fed but well nourished people.

Other than providing for the protein needs of much of the world population, however, the sea offers little promise in meeting the food requirements of our expanding populations. We must depend upon terrestrial agriculture for the energy rich foods such as cereals, grain, and rice. Those people who predict substantial contributions of plant food from the sea are deluding themselves and the public. It is true that photosynthesis and the production of plant material in the oceans equals or exceeds that of the land areas of our earth. We must understand, however, the enormous difference between the cycle of life in the sea and that on the land.

The plant production in the ocean is performed by microscopic organisms called the phytoplankton. When conditions are right these single-celled organisms can reproduce very rapidly and double or even triple the size of the population in a period of a single day. The phytoplankton produced are rapidly removed by the herbivorous zooplankton, however, so that at any given time the standing crop of plant material in the ocean is very small compared to the amount found on land. It is the rapid rate of turnover and conversion to zooplankton which accounts for the high rate of productivity by the low quantity present. Ryther (1960) for example, concluded that although the standing crop of phytoplankton of the sea constituted only one tenth of one percent of the total plant material on earth, the annual production by this small population is 40% of the total world annual production of plant materials. Thus the life cycle in the sea is measured in terms of days in contrast to the annual production of terrestrial crops which are replaced during a growing season or to the standing crop of living plant material in a forest which may take, on the average, 50 years to produce. To harvest the living plants of the sea it would be necessary to filter a million pounds of seawater to obtain one pound of plant material even when conditions are favorable for the growth of the phytoplankton. It is obvious that the cost and expended energy of obtaining this food would far exceed the energy cost of obtaining food from a land crop.

The grazing zooplankton of the sea concentrate this plant food into larger packages which are in turn suitable in size for consumption by the small fish wich in turn are the food for the larger organisms. Each step or trophic level, involves a loss of energy and food value since each organism requires part of the food it eats to maintain its own existence and to search for and capture additional food. The ratio of food eaten to new living material produced can be expressed as an efficiency which can be quite variable in the various steps of this food chain or trophic level. Zooplankton efficiencies as high as 40% have been recorded; the small rapidly growing fish may have efficiencies in the 20–30% range, but the efficiencies decrease as the organism becomes larger and more mature. Thus a large adult fish of nearly maximum size is utilizing virtually all of the food it eats to maintain its own existence.

It is obvious that each additional trophic level in the sequence of life leading to the product harvested decreases the size of the available harvest. Today we are harvesting only about 0.03% of the annual productivity of the microscopic plants of the oceans. About 40% of this harvest consists of small fishes such as sardines, anchovies and herring or of molluscs or crustaceans. These organisms live on a mixed diet of phytoplankton and herbivorous zooplankton, and we can consequently expect to achieve about 5% of the organic material produced by the plants in the sea by harvesting these forms. The enormous fisheries of Peru, for example, are based upon the anchovy which is very effective in 'biological efficiency'. In contrast, the fisheries on the banks of the North Atlantic, such as George's Bank and the Grand Banks are mostly at the second carnivore stage so that the sequence is somewhat as follows:

1000 pounds of phytoplankton produces
100 pounds of zooplankton or shellfish
50 pounds of anchovies and other small fishes
10 pounds of the smaller carnivores
1 pound of the carnivore harvested by man.

We might double or even quadruple the harvest from the sea but we cannot expect tremendous increases so long as we depend upon this inherently inefficient biological transfer by the natural system (Schaefer and Alverson, 1968). Some promise may be held by developing aquaculture and, indeed, properly managed fish farms can produce more protein per acre than properly managed farmland. In terms of solving the hunger problems of the world, however, it is only as a source of protein that the sea holds forth much promise.

I have described the cycle of life in the sea in some detail in order to conclude with some remarks about the possible effects of pollution upon this cycle. It is clear that there are many steps involved and serious interference with any one of these steps could adversely affect the harvest from the sea. For example increasing the turbidity of estuaries by increasing the silt load in a river will seriously diminish the photosynthetic production of organic matter by the phytoplankton, the first and most essential step in this food chain. All of the herbivorous zooplankton pass through several larval stages, many of which are more susceptible than the adults to environmental stress. Interference at any given larval stage could seriously effect the amount of final product which could be produced.

As mentioned earlier, drastic pollution of our estuaries can reduce not only the populations that we normally harvest there but can also interfere with the breeding cycles of many marine species. So far the pollution of the high seas has been identified but no direct effect on the total production of living material has been demonstrated there. We now know enough, however, about the cycle of life in the sea so that we should be able to evaluate possible effects and to prevent irreparable damage before it happens.

I hope that this paper has convinced you that the living resources of the sea are of considerable value to mankind and I hope also that man is wise enough to control his activities so that he can preserve and utilize wisely the resources that he needs to maintain his own life upon this planet.

References

Ryther, J. H.: 1960, 'Organic production by Plankton Algae, and Its Environmental Control', in *Ecology of Algae*, The Pymatuning Symposium in Ecology 18–19 June, 1959. Spec. Publ. Univ. Pittsburgh Press, No. 2, pp. 72–83.

Schaefer, M. B. and Alverson, D. L.: 1968, 'World Fish Potentials', in *The Future Fishing Industry of the United States*, University of Washington Publications in Fisheries, New Series, vol. 4, pp. 81–85.

For Further Reading

1. For general information in a popular vein:
   Wesley Marx, *The Frail Ocean*, Ballantine Books, New York, 1967, 274 pages.
2. On the social, economic, and strategic value of the ocean:
   E. A. Gullion (ed.), *Uses of the Sea*, The American Assembly, Prentice-Hall, Inc., Englewood Cliffs, N.J., 1968, 202 pages.
3. For essays by various specialists in the field:
   T. A. Olson and F. J. Burgess (eds.), *Pollution and Marine Ecology*, Interscience Publishers, John Wiley and Sons, New York, 1967, 364 pages.
4. For an introduction to the technical literature:
   E. A. Pearson (ed.), *Advances in Water Pollution Research*, Volume 3, Pergamon Press Ltd., London, 1964, 437 pages.

INTERACTIONS BETWEEN OCEANS AND TERRESTRIAL ECOSYSTEMS

BENGT LUNDHOLM

Swedish Ecological Research Committee,
Natural Science Research Council, Stockholm, Sweden

Abstract. The interactions between the different ecosystems are of importance, when pollution problems are penetrated. Even if oceans often are the sink for many pollutants, it is, however, possible that a transport takes place from the sea to the air. This material is later washed out over the continents. DDT may assemble in the thin oilslicks on the oceans and may be airborne in the breaker zones along the coasts. Also sulphur may leave the marine environment and by forming sulphuric acid contribute to the acidity in the rainwater.

In Sweden the interest is focused on two main pollution problems: the mercury situation and the acid rains. The lesson learnt is, that the pollutants move freely between air, water, soil and living organisms. There are also indications that these two pollution complexes are interlocked. The total environment must always be considered and the global pollution dynamics is of utmost importance.

1. Introduction

The oceans are often regarded as the final sink for many pollutants, but the toxic wastes may return to man in his food products. It is, however, possible that we also have other important influences from the sea on the terrestrial ecosystems. I want to give two examples, which may illustrate how the pollutants of the sea affect the terrestrial conditions.

DDT and its metabolites are assembled in the oceans, transported by rivers and winds. It has – as Dr. Goldberg mentioned – (p. 178) been found that African pesticides appear in the Caribbean winds. When this DDT reaches the aquatic environment, things happen very quickly. It has been shown that DDT added to an algal culture will be absorbed by the cells in less than 15 seconds. The lipid soluble DDT passes quickly through the cell wall into cells (Sodergren, 1968). (The solubility of DDT in water is extremely low – 1.2 μg/l – while its solubility in lipids is in the order of 100 g/l). The DDT is thus trapped in the organic material and is concentrated in the lipids, either in plants or animals. What happens when the cells are broken down? It has been noticed that very often areas with thin films of lipid materials appear on the sea surface. These slicks, which originate from broken-down cells may also contain the fat soluble pesticides. These films are broken up and the material is airborne especially in the breaker zones along the coasts. Dr. Erik Eriksson at the International Meteorological Institute in Stockholm has told me about this mechanism. He estimates that of the particles found in the air, 50% is organic material and half of this is fat or lipids, mostly transported from the sea. Even if it is not proved, the possibility of a transport of DDT in the same way seems likely.

In Europe, chlorinated hydrocarbons transported by rain have only been analyzed in Great Britain. It is rather remarkable that, during winter (November–January), the two coastal stations have an increased amount of DDT metabolities and higher amounts than the inland stations (Tarrant and Tatton, 1968).

Another interesting detail is mentioned in the paper about: 'Pesticides: Transatlantic Movements in the North East Trades', written in *Science* **159** (1968) by Dr. Goldberg, among others. The landward winds at La Jolla have a high amount of pesticides, but, unfortunately, we here have "an unknown admixture of air from neighboring agricultural areas". The theory about marine origin of some DDT may be very easy to test at La Jolla by analyzing the contents from different wind directions separately.

2. Marine Eutrophication

The other example of interactions between a marine and a terrestrial environment is from the Baltic. I have some graphs illustrating the situation. We have a very marked trend in the Baltic towards decreased oxygen and increased phosphorus in the deeper parts (Figure 1). We see on the graph that H_2S has appeared this year. These changes

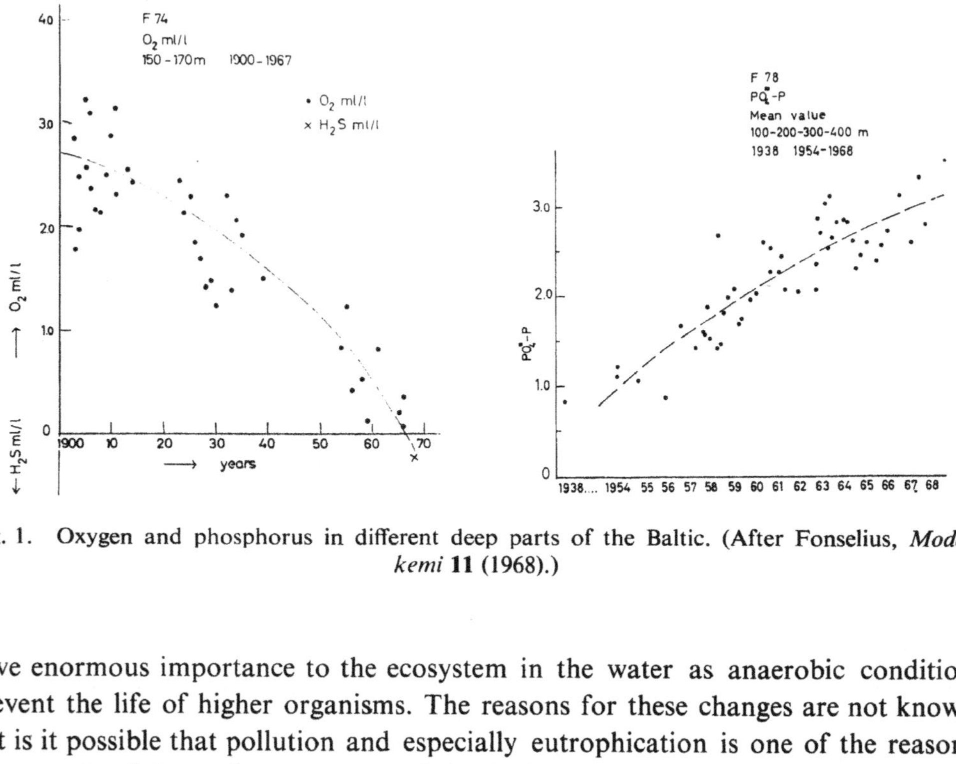

Fig. 1. Oxygen and phosphorus in different deep parts of the Baltic. (After Fonselius, *Modern kemi* **11** (1968).)

have enormous importance to the ecosystem in the water as anaerobic conditions prevent the life of higher organisms. The reasons for these changes are not known, but is it possible that pollution and especially eutrophication is one of the reasons. As a result of these changes, some of the hydrogen sulphide may leave the water; in the atmosphere, it can be oxidized to sulphuric acid. When this acid is washed out by the rains, the precipitation will be more acid with a lowered pH. In Sweden we have recorded a very marked increase of the acidity in the rains the last 10 years (see Figure 2). The main reason for this is, of course, the increased industrial air pollution by SO_2, but we are now also discussing other sources such as H_2S from

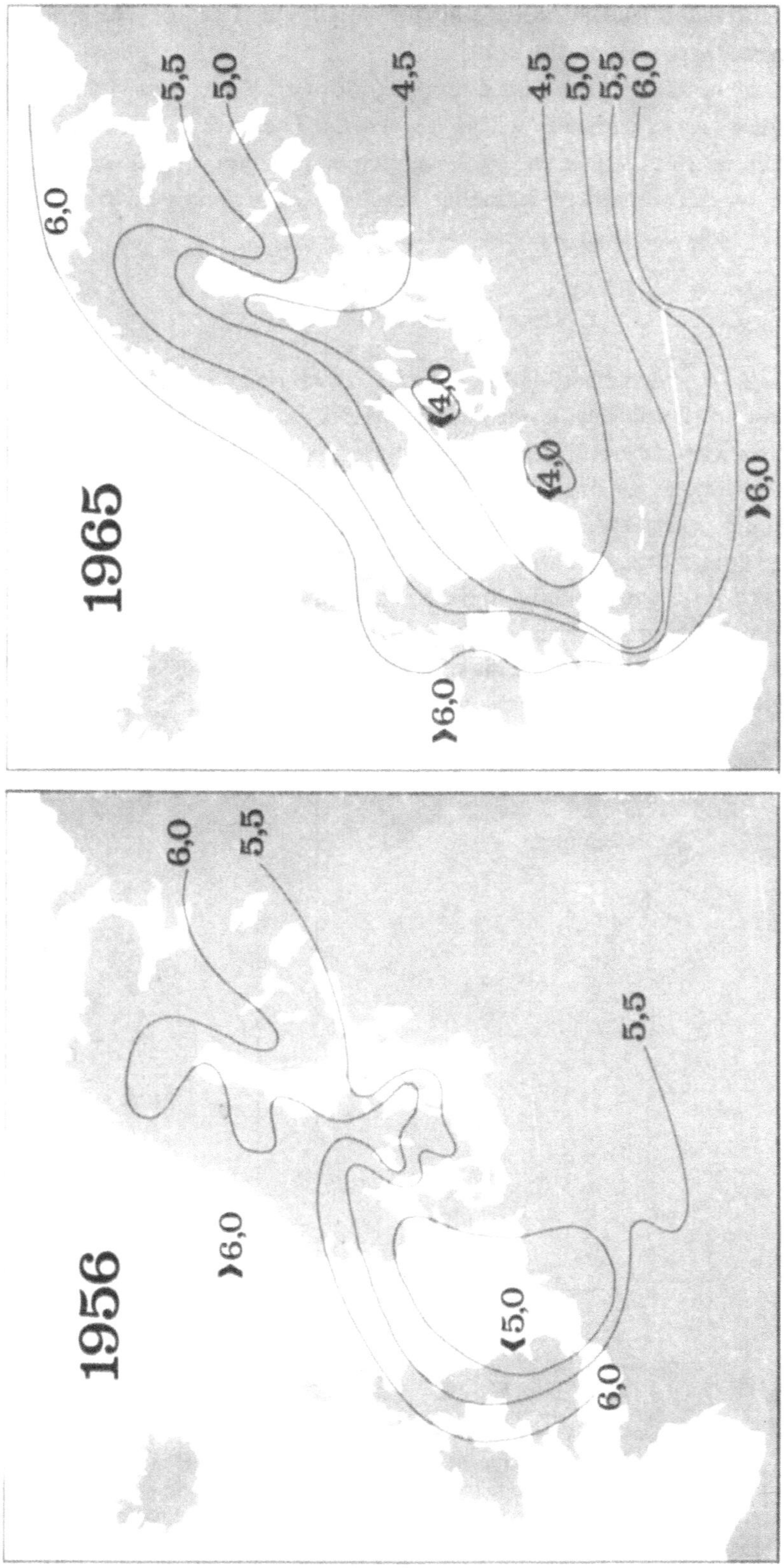

Fig. 2. pH in precipitation over Europe. The annual values for pH were obtained from monthly observations from the network of atmospheric chemistry stations. All data are available from the International Meteorological Institute, Tulegatan 41, Stockholm. (From S. Odén, *Forskning och Framsteg* **1** (1969).)

polluted marine areas, both the Baltic and the North Sea. This may be a problem in all polluted coastal areas along the continents.

Even if the acid rains have nothing directly to do with the oceans, I think I would like to follow the chain of causality with a few words. The acid rains affect the surface waters, especially in areas where the buffering capacity is low in the water. In oligotrophic waters, we get a marked influence resulting in a decreased pH, which has been recorded in large areas in western Sweden.

3. Heavy Metals Pollution

These events might be connected with another Swedish pollution problem; the occurrence of mercury in the natural environment. Mercury has been used as a pesticide and in connection with different industrial activities. We are especially concerned with the aquatic environment as both lakes and costal areas have been blacklisted and commercial fishing is forbidden, as the mercury content in the fish is too high. We have now found that mercury or any mercury compound, poured out in the water is transferred to the very toxic methyl mercury compounds. This methylation is performed by microorganisms in the sediments. A monomethyl compound (CH_3Hg) is formed in the water and in an alkaline environment this is transferred to a dimethyl compound $(CH_3)_2Hg$, which leaves the water as it is very volatile. By these mecha-

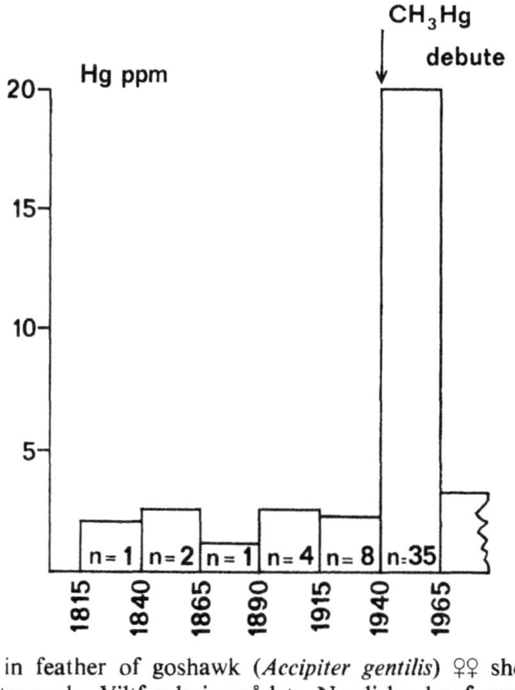

Fig. 3. Mercury levels in feather of goshawk (*Accipiter gentilis*) ♀♀ shot in May–June. (After A. Johnels and T. Westermark, Viltforskningsrådets Nordiska konferens 3–5/3 1966 and with results from the contribution by T. Westermark *et al.* at the symposium on mercury arranged by Nordforsk at Lidingö 10–11.10 1968) *n* refers to size of sample and 'CH_3Hg debute' marks the introduction of these compounds for seed dressing.

nisms, the mercury is removed from the sediments to the atmosphere. In an acid environment, the dimethyl compound is not formed and the mercury is trapped in the water as a monomethyl compound and is available to the ecosystems. An increased acidity in the surface waters will thus give a higher level of mercury. As a result of this, it is possible to find high mercury contents in fishes in unpolluted lakes very remote from both agricultural and industrial activities, especially as the biomass in these lakes is small and the amount of available mercury is fairly high. The mercury may have been transported to these lakes by air. As a result of these high mercury levels in fresh water and fresh water ecosystems and by the runoff to the sea, the coastal marine areas are also polluted by mercury. And the pollution circle is closed.

The problems in Sweden began with methylmercury compounds used in seed dressing. This resulted in very high mercury levels in terrestrial food chains. In 1966, this compound was forbidden in Sweden and now we can with satisfaction record that the mercury levels have dropped (see Figure 3). In aquatic environments, however, we still have very high mercury levels(see Figure 4). Here the sources are

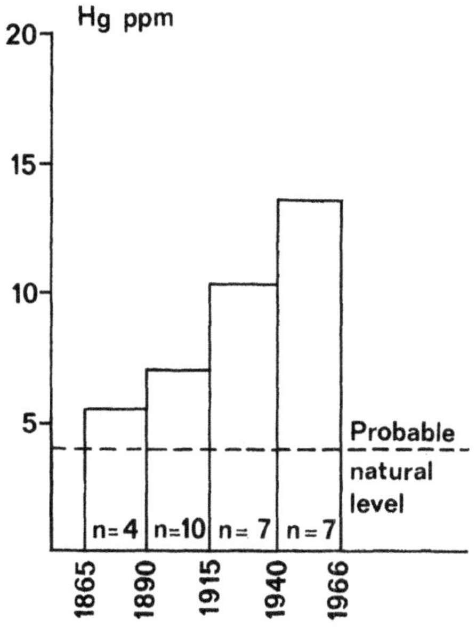

Fig. 4. Mercury levels in feather of great crested grebe (*Podiceps cristatus*). (Redrawn from A. Johnels and T. Westermark, Viltforskningsrådets Nordiska konferens 3–5/3 1966.)

different and not the same as in the terrestrial ecosystems. The increase of the levels started much earlier than in the terrestrial food chains. These high mercury levels especially in fish from fresh waters and some polluted coastal waters have caused trouble, as the Swedish government has been forced to blacklist several areas where fish is not regarded as suitable for human consumption.

In these two cases, birds of prey have been used as 'indicator species' to the accumulation of mercury. In Sweden, we have also tried to evaluate the importance of the increased outflow of lead into the environment. We have tried to find lead accumulation in birds of prey and also predatory fish, but we have so far not succeeded in getting a clear picture of the situation. Recently, however, certain species of mosses were shown to be good indicators for lead accumulation (see Figure 5). I think it is very important to realize that the different species react very differently to the same pollutant and that we have to find specific indicators to every pollutant.

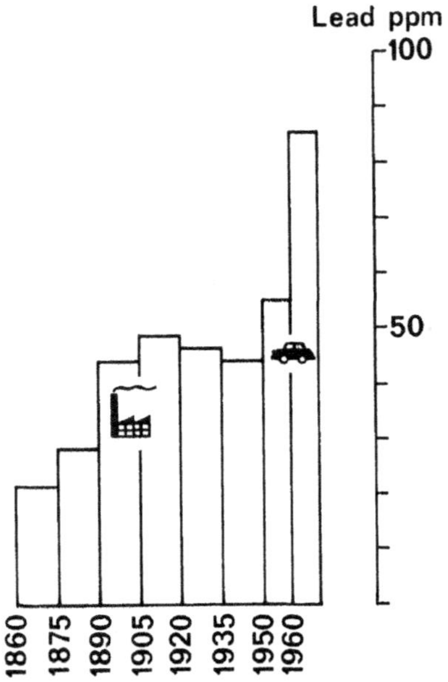

Fig. 5. Lead content in three different species of mosses (*Hylocomium, Pleurozium and Hypnum*). (After Å. Rühling and G. Tyler, *Botaniska Notiser* **3** (1968).)

4. Pollution from Pesticides

Toxic wastes – such as pesticides – transported by the atmosphere are falling out in rains over both the oceans and the continents. On the continents, the wastes are accumulated in soils. Depending on the climatic conditions, some of the substance is removed from the soil. This transportation may be measured by the removal time for half of the amount. I prefer the term 'half removal time' to 'half life' as some of the compounds are still very much 'alive' after their removal from the soil.

These differences in removal time result in an increased accumulation in the northern soils. This has been shown for Swedish soils by Svante Oden at the University of Agriculture in Uppsala (see Figure 6). It would be of great interest to find out if we have similar latitude differences in the seas. Is this the explanation to the occurrence of DDT in the penguins? We now have to focus our scientific investigations on the global transport systems.

The marine ecosystems are extremely useful if we want to study the impact of pollution on the biosphere. In the oceans we find an integration of the large-scale

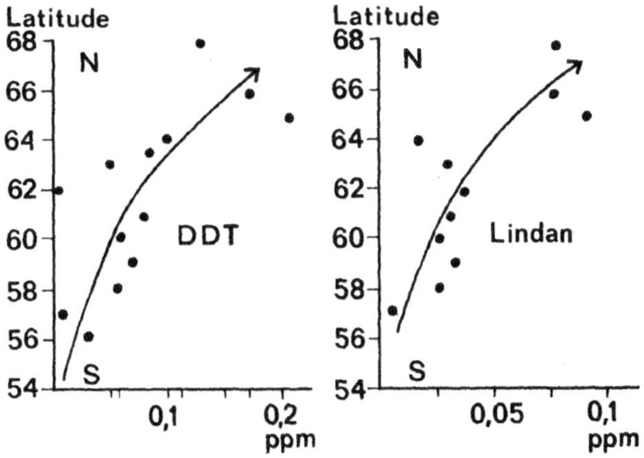

Fig. 6. Residue contents from north to south in Swedish soils, which have not been treated with pesticides. The material is based on 500 soil samples. (After S. Odén, *Handelstidningen* 21.10.1968.)

processes. There are no local disturbances which mostly complicate the studies of terrestrial ecosystems. By using the oceans as study objects, many basic problems concerning global pollution may be solved. To take one example, DDT and other persistent pesticides are man-made and new to nature. These substances can now be used as labelled compounds. Their geographic distribution in the seas, their position in the marine ecosystems, and their transport ways to and from the oceans will give us a picture of the global pollution.

The international status of the seas makes them easily available to scientific investigators. By using existing resources and facilities for scientific research, it would not be too difficult to plan and carry out a model study of marine pollution. Even with rather limited funds it would be possible to reach outstanding results in a short time.

If we start immediately with the investigations, it would be possible to get fairly good ideas about the general situation concerning global pollution in time for the United Nation's conference about the human environment in 1972. At this conference, the scientists will be able to state their case to the politicians.

The urgency of the situation, or rather the urgency of getting facts which allow us to evaluate the situation, is a challenge to the scientific community. I also think this is an area for cooperation between scientists from different nations, but also an area when even just a few cooperating nations could make valuable progress.

References

Goldberg, E. D. *et al.*: 1968, 'Pesticides, Transatlantic Movements in the North East Trades', *Science* **159**, 1233-1235.

Sodergren, A.: 1968, *Oikos* **19**, 126–138.

Tarrant, K. R. and Tatton, J. O'G.: 1968, *Nature* **219**, 725–727.

SOME UNSOLVED PROBLEMS: A PANEL DISCUSSION

Following the prepared presentations and discussions of the Symposium, there was held a panel discussion chaired by A. F. Spilhaus (President-Elect of the American Association for the Advancement of Science). Panelists were J. L. Buckley (Office of Science and Technology), W. F. Libby (University of California, Los Angeles), B. Lundholm (Swedish Ecological Research Commission), R. Revelle (Harvard University), W. O. Roberts (University Corporation for Atmospheric Research), and S. F. Singer (Department of the Interior).

Many detailed questions were considered by the panel but not always settled conclusively. Some examples are:

Can natural processes for removing CO_2 from the atmosphere be stimulated or otherwise affected?

What are the main contributors to the regeneration of atmospheric oxygen?

Can the burning of fossil fuel, even of all the reserves, have an appreciable effect on atmospheric oxygen content? It was computed that the decrease would be less than one-half of 1%.

How does atmospheric turbidity induced by human activity compare with natural turbidity? What is the projection for the future? Are we in danger of triggering a feedback mechanism which will produce irreversible and possibly catastrophic consequences? What is the effect of jet planes on turbidity? What effect do high-flying jet planes have on the chemistry of the upper atmosphere, on water vapor and ozone, and on the radiative properties of the upper atmosphere?

What is the effect on climate due to the competing influences of the increased albedo from high altitude cirrus, the increased thermal emission, and the increased 'greenhouse' effect? Where are the areas where human activities can induce major changes? One example mentioned is soot which falls on snow and greatly accelerates melting.

In connection with the nitrogen problem, the question was raised concerning the relative contribution to atmospheric nitrogen compounds from excessive use of fertilizers, from motor cars, and from natural causes such as lightning. It was brought out that the real cost of eutrophication is not borne by the food grower, but is a distributed social cost. It was concluded that drastic changes are necessary in soil management in order to eliminate the fertilizer pollution problem and that this would involve economic, political, and moral issues.

With regard to ocean pollution, there was a further discussion of the methods of transport in the global atmosphere-ocean system. For example, evaporation of gasoline and dry-cleaning compounds introduces hydrocarbons into the atmosphere which are then washed into the ocean. In addition to these and other inadvertent discharges, there are, of course, the direct discharges from the outfalls of coastal cities and from ships. No public policy had as yet been announced concerning the degree of treatment which would be appropriate for such discharges of wastes.

Some of the scientific questions are still unsolved: e.g., what is the fate of viruses that are discharged into sea water; how are they deactivated? What is the lifetime of DDT in the ocean; how is it broken down? What are the effects on marine organisms of exotic chemicals, many with specific biotic properties? And, finally, since the oceans can transport materials over nearly all of the earth's surface, what are the possible long-range effects on atmosphere, climate, and man which can arise from ocean pollution?

There was general agreement on a few matters. Too few people are working on these vital problems and, in many cases, only on a part-time basis. There is increasing competition for resources held in common and, therefore, much of the basic problems are economic in nature. Since half the world's population depends on marine proteins, we have an obligation to protect the marine environment. What we really need is a policy on population since it is the pressure of population which, in turn, creates the pressure on natural resources and on environmental quality.

This sampling of problems makes it evident that many of the questions are very far-reaching but, unfortunately, also very difficult. Yet the subject of global effects of environmental pollution is of such importance to the world as a whole that we cannot relax our efforts to settle the scientific as well as the economic and political problems. The Symposium provided a good curtain raiser for worldwide efforts to set up a global monitoring system of crucial pollution parameters, as well as for the forthcoming United Nations Conference of 1972 dealing with the Problems of the Human Environment.

EPILOGUE

The oil accident off Santa Barbara offers an interesting study in public reaction. It was massive, unexpected, and produced consequences which were quite apparent to the general public. And it lent great emphasis to hearings on oil pollution legislation which were just then before the Congress. By contrast, the far-reaching pollution of our total global environment by various kinds of poisons has produced no such dramatic effects and certainly not any equivalent amount of public reaction.

The oil accident produced primarily local effects, mainly damaging the recreational value of some 30 miles of beaches. The oil itself, while toxic to various marine organisms, especially in its aromatic fractions, is degraded by bacteria and has a relatively short lifetime. It is the intense concentration in space and time which accounts for the seriousness of this oil spill of some 10^{10} g and of the *Torrey Canyon* accident of 10^{11} g. In addition, there is a yearly worldwide spillage of about 10^{12} g, as well as natural oil seeps; there are hydrocarbons produced by various marine organisms, as well as hydrocarbons evaporated from the land (10^{13} g of gasoline, 10^{12} g of various solvents); and some 10^{12} g of waste motor oil per year are dumped in the U.S.A. alone.

Consider, in contrast, the chlorinated hydrocarbons released into the environment; the insecticides DDT, dieldrin, and related chemicals, or any of the polychlorinated biphenyls used in the manufacture of plastics, paints and rubber. These substances are not specific; they are toxic to many organisms. They are extremely persistent. They dissolve in lipids and therefore concentrate in the fatty tissues of organisms. As a result, the upper members of the food chain now carry large concentrations of DDT or its breakdown products which presumably affect their hormone metabolism. DDT, for example, stimulates the production of hepatic enzymes in mammals, which interferes with the action of certain drugs and also of steroid hormones such as the estrogen or testosterone produced by the animal. In rats, increased enzyme activity occurred at a concentration of 10 ppm in fatty tissue. The average human in the U.S.A. now stores about 12 ppm of DDT and DDE in his fat – making him unfit for human consumption!

The most precise evidence on DDT effects relates to carnivorous birds. What we are facing now is the extinction of many species of seabirds because of lack of success in breeding, and perhaps of other marine animals, as the concentration of DDT in the ocean grows. The irony is that there are now available insecticides which are quite specific in their application to target organisms and others which degrade rapidly in the environment. But how do we carry the message?

Mankind has a record of reacting *after* a disaster strikes. Dams are built after floods, not before. So far in human history, disasters have not taken place on a global scale. Therefore we don't really have a tested mechanism for dealing with global threats, such as a long-range, worldwide degradation of the environment. If we ignore the present warning signs and wait for an ecological disaster to strike, it will probably be too late.

The distinguished biologist Garrett Hardin has pointed out how very difficult it is psychologically to really believe that a disaster is impending. "How can one believe in something – particularly an unpleasant something – that has never happened before?" This must have been a terrible problem for Noah. Can't we just hear his complacent compatriots: "Something has always happened to save us." or "Don't worry about the rising waters, Noah; our advanced technology will surely discover a substitute for breathing." Unfortunately, the Bible doesn't tell us much about Noah's psychological trials and tribulations. But if it was wisdom that enabled Noah to believe in the 'never-yet-happened', we could use some of that wisdom now.

S. FRED SINGER

Deputy Assistant Secretary
U.S. Department of the Interior
Washington, D.C., U.S.A.

January 1970

ABOUT THE AUTHORS

FRANCIS S. JOHNSON is Acting President of the University of Texas in Dallas, Texas. He received his Ph.D. in meteorology from the University of California, Los Angeles, and has done extensive research in meteorology and space physics, with particular emphasis on the upper atmosphere. He is a member of numerous committees of the National Academy of Sciences, including the Space Science Board and the Advisory Committee to the Environmental Science Services Administration. He is a member of the Executive Committee of the International Association of Geomagnetism and Aeronomy, and the English Secretary of Commission IV of the International Union of Radio Science. He has served as Associate Editor of the *Journal of Geophysical Research*. His most recent research interests include the development of planetary atmospheres.

FREDERICK D. SISLER is a marine microbiologist who received his Ph.D. from the Scripps Institution of Oceanography in La Jolla, Calif., in 1949. He has held positions in private industry and government laboratories and is presently employed by the Federal Water Pollution Control Administration of the Department of the Interior. His original research has included such topics as biochemical fuel cells, organic matter in meteorites, and oceanographic instrumentation.

SYUKURO MANABE received his Ph.D. in meteorology from Tokyo University in 1958, and since then has served as Research Meteorologist with the U.S. Weather Bureau. Since 1962 he has been a member of the Geophysical Fluid Dynamics Laboratory, now located in Princeton, N. J., and concerned with the simulation of atmospheric circulation problems by means of high speed electronic computers.

WILLARD F. LIBBY received degrees in chemistry from the University of California, Berkeley, and in addition holds a number of honorary doctorates. He teaches at the University of California, Los Angeles, where he is Director of the Institute of Geophysics, and at the University of Colorado. He has served a five year-term as Commissioner of the U.S. Atomic Energy Commission. He is a Nobel Laureate in Chemistry (1960) in recognition of his work on natural radiocarbon and its application to the dating of ancient artifacts. He has received many honors and awards, including the Willard Gibbs Medal of the American Chemical Society, the Day Medal of the Geological Society of America, and the Albert Einstein Medal. He has served on numerous scientific advisory groups, as a Consultant and Director of many industrial concerns, and is affiliated with scientific societies both in the U.S.A. and abroad.

RAINER BERGER received his Ph.D. in organic isotope chemistry at the University of Illinois in 1960. He is Associate Professor of Geophysics and History at the University of California, Los Angeles, having been affiliated with aerospace firms in California. His interests now include the application of radiocarbon dating to archeology and anthropology.

LOUIS S. JAFFE is a chemist and physiologist with degrees from Boston College and Columbia University. He is currently a consultant in the areas of environmental health, pollution and related fields. For the past 30 years he has held various responsible government positions in these areas. More recently he was concerned with developing air quality criteria and standards for the Department of Health, Education and Welfare. Before that, he served as a toxicologist concerned with environmental protection and flight safety for the Federal Aviation Administration and before that held various environmental protection positions in the Department of Defense. He is the author of many publications concerning environmental health and atmospheric pollution and has been particularly concerned with the problems of ozone and carbon monoxide.

ELMER ROBINSON, Senior Meteorologist and Chairman, Environmental Research Department, Stanford Research Institute. Mr. Robinson graduated from the University of California at Los Angeles in 1948 with an M.A. degree in Meteorology. In 1948 he joined Stanford Research Institute and has been active in research programs dealing with air pollution, cloud and fog physics, and atmospheric chemistry. During the period 1957–1960 he served as Chief, Air Analysis Section, Bay Area Air Pollution Control District, San Francisco, California.

ROBERT C. ROBBINS, Senior Physical Chemist, Environmental Research Department, Stanford Research Institute. Dr. Robbins received his Ph.D. from the University of Delaware in Physical Chemistry in 1953. Since 1954 he has been active in various fields of atmospheric research at Stanford Research Institute. Dr. Robbins special interests have included air pollution photochemistry, atmospheric chemistry, aerosol chemistry and physics, and atmospheric carbon monoxide research. Dr. Robbins served as a meteorologist with the U. S. Navy in WW II and has also been with the du Pont Co.

BARRY COMMONER received his Ph.D. from Harvard University in 1941. Following military service, he joined the faculty of Washington University in St. Louis, becoming Chairman of the Department of Botany in 1965 and Director of the Center for the Biology of Natural Systems. He has received many honors including honorary degrees and the Newcomb Cleveland Prize of the American Association for the Advancement of Science. He has served as Chairman of the AAAS Committee on Science in the Promotion of Human Welfare and, in 1967, was elected to the AAAS Board of Directors. He has been actively investigating fundamental problems relating to the physio-chemical basis of biological processes, including pioneer studies on free radicals and on virus replication. He has also developed a deep interest in the interaction between science and social problems; his book *Science and Survival* deals with the threats to human survival from technological changes.

D. R. KEENEY is associate professor, Department of Soil Science, University of Wisconsin, Madison. He received his B.S. (1959) from Iowa State University, his M.S. (1961) from the University of Wisconsin, and his Ph.D. (1965) from Iowa State University. He joined the University of Wisconsin staff in 1966 after a year's postdoctorate study at Iowa State. Dr. Keeney's research deals primarily with methods for determining various forms of nitrogen and with nitrogen transformations in soils, lake sediments and waters.

W. R. GARDNER is professor, Department of Soil Science, University of Wisconsin, Madison. He received his B.S. (1949) from Utah State University and his M.S. (1951) and Ph.D. (1953) from Iowa State University. Before joining the University of Wisconsin staff in 1966, he served for 15 years as physicist, Agricultural Research Service, U.S. Department of Agriculture, Salinity Laboratory, Riverside, Calif. He has pioneered investigations on water and solute movement through soils and the ion transport phenomena in plants. He also is actively conducting research on micrometeorology, hydrology, and the movement of pollutants through soil.

THEODORE D. BYERLY is Assistant Director of Science and Education of the U.S. Department of Agriculture. He received his Ph.D. from the University of Iowa in 1926 and has been associated with universities and agricultural research throughout his career. He has received many honors and awards including the Department of Agriculture Distinguished Service Award. He served as Chairman of the Division of Biology and Agriculture of the National Academy of Sciences/National Research Council.

ARTHUR D. HASLER received his Ph.D. from the University of Wisconsin in 1937 and is associated with that institution as Professor of Zoology and Director of the Laboratory of Limnology. He has been the recipient of many honors including election to the National Academy of Sciences, honorary doctorates, and President of the American Society of Limnology and Oceanography. He is also a member of several Finnish scientific societies. He serves on many international scientific organizations and was Chairman of the XVth International Congress of Limnology and Oceanography.

REID A. BRYSON received his Ph.D. from the University of Chicago in meteorology and has taught mainly at the University of Wisconsin, where he established the Department of Meteorology and now serves as Professor of Meteorology and Geography. He has received many honors including

Honorary Vice President of the International Quaternary Association. His special interest has been world climatology, and he has several years of field experience in many parts of the world.

WAYNE M. WENDLAND received his M.S. at the University of Wisconsin in 1965 and is a Ph.D. candidate in meteorology and Instructor in Geography there at present. He has been a Weather Forecaster and Weather Detachment Commander in the U.S. Air Force where he served for 8 years. His areas of interest are historical climatology and the relation between climate and man.

J. MURRAY MITCHELL received his Ph.D. in Meteorology at the University of Pennsylvania in 1960. He has been Research Meteorologist with U.S. Weather Bureau since 1955. He has been appointed as a Visiting Lecturer at a number of institutions, is a member of the U.S. National Committee, International Association for Quaternary Research, and a member of committees of the American Meteorological Society and American Geophysical Union concerned with paleoclimatology and climate change. He has served as Associate Editor of the *Journal of Applied Meteorology* and is presently Editor of the Meteorological Monograph Series.

VINCENT J. SCHAEFER is Director of the Atmospheric Science Center of the State University of New York at Albany. For most of his career he has been associated with the General Electric Research Laboratory and principally with Dr. Irving Langmuir, with whom he is co-inventor of the artificial-fog smoke-screen generator. He is also the discoverer of the dry ice seeding technique for cloud modification. He has received an honorary doctorate from the University of Notre Dame and the Robert M. Losey Award of the Institute of Aeronautical Sciences.

EDWARD D. GOLDBERG received his Ph.D. in chemistry from the University of Chicago in 1949. He then joined the Scripps Institution of Oceanography where he now serves as full Professor. His scientific interests include radio chemistry of marine waters and sediments but extend into problems of planetary atmospheres. He is active in professional societies and Editor of three journals. He was elected Vice President of the Section of Volcanology in the American Geophysical Union.

GEORGE M. WOODWELL received his Ph.D. from Duke University in 1958. He now serves as Senior Ecologist in the Biology Department of the Brookhaven National Laboratory on Long Island, New York. In addition, he holds an adjunct appointment as Lecturer in Ecology at Yale University. His research interests have focused on the ecological effects of ionizing radiation and on the movement of nutrient elements and toxic materials through the various biological, geological, and chemical cycles. His more recent interests include the cycling of DDT over the surface of the earth. His society memberships include not only scientific societies but also the Sierra Club and Nature Conservancy. He is a member of the Board of Trustees of the Environmental Defense Fund, designed to use the courts to restrict the use of persistent pesticides.

BOSTWICK H. KETCHUM received a Ph.D. in biology from Harvard University in 1938. Most of his professional career has been spent at the Woods Hole Oceanographic Institution. He is presently on leave at the National Science Foundation, serving as Section Head, Ecology and Systematic Biology. He is involved in many professional societies and in editorial activities. He serves as a Consultant to the U.S. Public Health Service and is a member of advisory committees of the National Academy of Sciences and other scientific organizations. His range of interest includes marine ecology, physiology of algae, nutrient cycles in the ocean, and the special problems of estuaries.

BENGT LUNDHOLM is Executive Member of the Ecological Research Committee at the Swedish Natural Science Research Council in Stockholm. He is also Chairman of the ad hoc Committee on the Human Environment which was set up by the International Council of Scientific Unions in 1968. When the International Biological Program created a special working group on global baseline stations, he was chosen as Chairman. He graduated from the University of Uppsala and spent 3 years in east and south Africa studying mammals. During 1964–68 he was Secretary-General in the Swedish Royal Commission on Natural Resources which laid the basis for the present Swedish policy concerning the environment.

S. FRED SINGER, who served as convenor of the symposium and editor of this volume, received his

Ph.D. in physics from Princeton University in 1948. He has alternated between universities and government service and now holds the position of Deputy Assistant Secretary, U.S. Department of the Interior. Following his earlier association with the U.S. Weather Bureau as Director of the National Weather Satellite Center, he received the Gold Medal Award for Distinguished Federal Service. He has served as Dean of Environmental Sciences at the University of Miami. His scientific interests have been in atmospheric physics, space physics, oceanography, and the early history and development of the earth-moon system. His current professional concerns include a wide range of water problems, including ecological and environmental concerns. He is Chairman of the Committee on Environmental Quality of the American Geophysical Union.

INDEX OF NAMES

INDEX OF SUBJECTS

GPSR Compliance

The European Union's (EU) General Product Safety Regulation (GPSR) is a set of rules that requires consumer products to be safe and our obligations to ensure this.

If you have any concerns about our products, you can contact us on ProductSafety@springernature.com

In case Publisher is established outside the EU, the EU authorized representative is:

Springer Nature Customer Service Center GmbH
Europaplatz 3
69115 Heidelberg, Germany

Batch number: 10824727

Printed by Printforce, the Netherlands